APPLIED MACHINE LEARNING FOR HEALTH AND FITNESS

A PRACTICAL GUIDE TO MACHINE LEARNING WITH DEEP VISION, SENSORS AND IOT

Kevin Ashley

Foreword by Phil Cheetham

Apress®

Applied Machine Learning for Health and Fitness: A Practical Guide to Machine Learning with Deep Vision, Sensors and IoT

Kevin Ashley
Belmont, CA, USA

ISBN-13 (pbk): 978-1-4842-5771-5   ISBN-13 (electronic): 978-1-4842-5772-2
https://doi.org/10.1007/978-1-4842-5772-2

Copyright © 2020 by Kevin Ashley

Managing Director, Apress Media LLC: Welmoed Spahr
Acquisitions Editor: Natalie Pao
Development Editor: James Markham
Coordinating Editor: Jessica Vakili

Distributed to the book trade worldwide by Springer Science+Business Media New York, 233 Spring Street, 6th Floor, New York, NY 10013. Phone 1-800-SPRINGER, fax (201) 348-4505, e-mail orders-ny@springer-sbm.com, or visit www.springeronline.com. Apress Media, LLC is a California LLC and the sole member (owner) is Springer Science + Business Media Finance Inc (SSBM Finance Inc). SSBM Finance Inc is a **Delaware** corporation.

For information on translations, please e-mail booktranslations@springernature.com; for reprint, paperback, or audio rights, please e-mail bookpermissions@springernature.com.

Apress titles may be purchased in bulk for academic, corporate, or promotional use. eBook versions and licenses are also available for most titles. For more information, reference our Print and eBook Bulk Sales web page at http://www.apress.com/bulk-sales.

Any source code or other supplementary material referenced by the author in this book is available to readers on GitHub via the book's product page, located at www.apress.com/978-1-4842-5771-5. For more detailed information, please visit http://www.apress.com/source-code.

Printed on acid-free paper

Contents

About the Author

 **Kevin Ashley** is Microsoft developer Hall of Fame engineer and author of popular sports, fitness, and gaming apps, Skype-featured bots, and cloud platforms such as Active Fitness and Winter Sports with several million users.

Kevin is a professional ski instructor, with a lifelong history of connecting technology and sports, including working with the US Olympic Team and sport organizations, partners, and startups in Silicon Valley and worldwide. He is a passionate technical speaker and founder of several startups and ventures, including mobile, sports, fitness, and cloud.

About the Technical Reviewers

Phil Cheetham is currently the senior sport technologist and biomechanist for the United States Olympic Committee (USOC) at the Olympic Training Center in Chula Vista, California. His mission is to acquire, develop, and implement technology to help improve athletes' performance, with the goal of winning medals at the Olympic Games. Phil is an Olympian in gymnastics for Australia and was also an elite diver. He also holds a PhD in sport biomechanics.

Tom Honeybone is a software engineer at Microsoft with a passionate interest in machine learning especially as the technology is applied to the media, entertainment, and communications industries.

Olga Vigdorovich is a data scientist, database engineer, a Microsoft Certified Professional, and an avid skier. She works with IoT sensor data analysis and builds data models for scalable cloud platforms based on Microsoft Azure. Olga contributed to several Microsoft Developer Magazine articles about AI and sensors and participated in Active Fitness development.

Mike Downey is Principal Architect working with sports technology and partnerships at Microsoft.

Jacob Spoelstra is Director of Data Science at Microsoft, specializing in neural networks, machine learning, and analytics.

Max Zilberman is Principal Software Engineering Manager at Microsoft with a history of leading multidiscipline teams.

Hang Zhang is Principal Data and Applied Scientist at Microsoft, focusing on big data, IoT, and predictive modeling.

Foreword to AI for Health and Fitness

By Phil Cheetham

Kevin has asked me to write the foreword to this groundbreaking book on AI in health and fitness, and I am honored to do so. I met Kevin a few years ago. He was helping our Team USA coaches and athletes by developing technology to measure their performance characteristics. As a sport technologist for the United States Olympic & Paralympic Committee, that was my job too. Since then we have collaborated on several sport technology projects and happily, we continue to do so.

As part of this foreword, I would like to indulge in some reminiscing and describe to you my journey through technology and sports and to their combination into sports technology. I have worked with many sports while at the USOPC, but two of the sports that I am most knowledgeable and passionate about are gymnastics and diving. I competed in both in high school in Sydney, Australia where I am from. As a youth I reached national level in both sports; in diving, a silver medal in 1-meter and 10-meter events at the junior national championships, and in gymnastics I was the all-around champion, also at the junior national championships. This was way back in the early 1970s. After I graduated high school and at the beginning of university, I realized I couldn't keep doing both diving and gymnastics at that level, plus pursue my degree in electrical engineering. I made a choice. I chose gymnastics. While pursuing my bachelor's degree in electrical engineering I simultaneously continued rigorous training in gymnastics. By my third year of university (junior year) I won the Australian senior all-around championship, and in the following year, 1976, I achieved one of my life's goals, I became an Olympian. I competed in the Olympic Games Montreal 1976. That was the experience of my life and will never be forgotten. Thinking about marching in the Opening Ceremony with the Australian team, and all the other Olympians, still sends chills up my spine!

On the technical side, I was always interested in electronics as a kid and was a member of my high school radio club. I built crystal sets, transistor radios and even a vacuum tube amplifier that powered a speaker, with input from the crystal set. I tinkered a lot. That interest led to me choosing electrical engineering at university and continued into my thesis project which was an electronic simulation of the cardiovascular system. Using an oscilloscope as a display, my circuit would simulate the systolic and diastolic pressure pulses that the heart produces. It was very crude, but it worked and helped me get an honors degree in electrical engineering from the University of New South Wales, in Sydney, in 1977.

At the end of the 1970s I came to the United States to attempt to qualify for my second Olympics, to be held in Moscow. I trained at Arizona State University (ASU) with coach Don Robinson and his incredible team. They were at the top echelon of NCAA gymnastics for many years. To train with them Coach Robinson appointed me an assistant coach and I enrolled in a master's degree in biomechanics. Now I could combine science with my sport! For my master's thesis I wrote my first motion analysis software program in Basic on a Tektronics 4052 computer. It had 64K of memory and two 5 1/4 floppy drives. It also had a beautiful vector graphics screen for nice animations. The software I wrote allowed me to manually digitize the motions of sports skills from film. I used this software to do a biomechanical analysis of vaulting for my thesis. Several of my teammates volunteered and we all performed several handspring front somersault vaults with black tape marking our joints, while being filmed at 300 frames per second. Using my software, I hand digitized (annotated) each body joint on each frame of film for each gymnast. This was a very tedious process, placing the cursor on each joint one by one, then clicking to the next frame and repeating for over 600 frames for each vault. When done, the data I collected allowed me to create a stick figure and animate each performance. From the data the program calculated the body center of gravity, and its velocity. With this information I was able to statistically determine which were the key performance indicators in the handspring front somersault vault. I was able to tell my teammates how to improve their performances, and, importantly to me, I earned my master's degree in biomechanics. It also put me on the path to developing several motion analysis systems during my career.

One of my professors at ASU, Dr. Dan Landers, was on the scientific advisory board of the USOPC and one spring break in the early 1980s, he took us to Colorado Springs on a field trip. There I met Dr. Charles Dillman, the director of the sports science lab. A few months later a job as the research engineer for the lab became available and I got it. Now retired from gymnastics competition I was lucky enough to be working at the U.S. Olympic & Paralympic Training Center. I worked on practical science projects to help Team USA athletes achieve success. For one of my projects I developed a motion capture system using VHS video instead of film. We used an IBM XT, with an image capture card, a VCR controller card and code that I wrote in 8088 assembly language and Basic. This was a breakthrough at the time since now we could produce the stick figures and data much more quickly and cheaply than from film. We didn't have to send the film away to get developed and at $100 a reel compared to the price of a VHS tape, we definitely saved both time and money. During the day we would video tape the athletes' skills, then at night we would manually digitize the body points and then the next day, show the coaches and athletes the data and stick figure animations. It was an exciting time, and this was pioneering technology.

My career continued along this path, and after several years at the USOPC, I left to form a company called Peak Performance Technologies. We developed various new motion analysis systems, manual and automatic tracking, with 2D and 3D analysis, all based on video. The systems were designed for biomechanics research and many biomechanics, and kinesiology programs, at universities purchased them. Our flagship product was a 3D analysis system that used multiple video cameras to capture data from reflective markers placed on the joints of the subject. This made tracking much faster than doing it manually by hand, however, it was still not totally automatic since if a camera lost track of a marker we would have to go back and manually correct it.

Frustrated by these limitations, my brother Steve and I left Peak and formed a new company called Skill Technologies. This was in the early 1990s. We had found a new exciting electromagnetic motion tracking technology. We wrote new software and again developed research systems for universities. The exciting thing about this new system was that you could put on the sensors and see the skeleton figure on the screen moving as you moved in real-time. I will always remember demonstrating it at an expo when a gentleman walked by, saw my brother and the skeleton moving synchronously, did a double take, came back and asked, "Can you do that for golf?" We said yes and developed a golf swing analysis system to his specifications. That connection led to us developing the original motion capture software for GolfTec which is now one of the largest golf swing training companies in the world. Over the next several years we refined our system, and in the early 2000s met the Titleist Performance Institute (TPI) founders, Greg Rose and Dave Phillips and customized our motion analysis system to produce a full-body, real-time motion analysis system specifically for golf called AMM 3D. We also branded it as TPI 3D. TPI is still using it today to analyze the swings of many PGA tour pros.

In the early 2010s I had the opportunity to again take a position working for the USOPC, at what was then the Olympic Training Center in Chula Vista, California. I took the position as senior sport technologist and worked specifically with track and field throws events. My job was to discover and implement technology that would help improve performance and avoid injury.

While based in Chula Vista, I continued studying for my PhD in the same field as my master's degree, exercise science with a focus on sport biomechanics. My goal was to finish the PhD before my sixtieth birthday, and I did, barely. My dissertation was on the golf swing. It is titled "The Relationship of Club Handle Twist Velocity to Selected Biomechanical Characteristics of the Golf Drive". Basically, I looked at how the twist velocity of the golf club handle (akin to club face closure rate) affected distance, accuracy, wrist angles and body posture during the swing. It was published in August 2014 and my sixtieth birthday was in October 2014. I just made it!

One of our success stories while working at the training center in Chula Vista was in the shot put event. We helped two different companies modify their radar technology and adapt it for measuring the release characteristics of the shot put. This proved to be a very valuable feedback tool giving us the release angle, velocity, height, and direction of the shot put immediately after the throw. All these values also determined exactly what the throw distance would be. It saved us measuring the distance with a tape measure, but also told us exactly how steep or flat the throw was, how high they released the shot put, and how fast it came out of their hand. They now had benchmarks to achieve on each throw and in each session. Ryan Crouser and Joe Kovacs, both used the system regularly at training, and at the Rio Olympics in 2016 they won the gold and the silver medals, respectively.

That brings me pretty much up to date and to the time when I met Kevin. Kevin was visiting the training center with co-workers and we talked about how we could use sensor technology, specifically inertial measurement units (IMUs) to measure the important variables that determine the success of a performance. I introduced Kevin to Cyrus Hostetler a javelin thrower and an Olympian. Cyrus is very technically inclined and asked if Kevin could develop a sensor to measure the release characteristics of the javelin throw, the same as we were doing with the shot put. The radar was unable to do that accurately for the javelin. The javelin is too thin for the radar to track accurately, at least at that time. Kevin developed a sensor that fit onto the javelin. The sensor was imbedded in a 3D printed plastic ring that Kevin manufactured. A very creative solution.

My relationship with Kevin continued and next we collaborated on a project for diving; a project funded by the U.S. Olympic and Paralympic Foundation's Technology and Innovation Fund. This fund is sponsored by donors and includes several high-level technology executives from Silicon Valley. Its goal is to maximize the impact and utilization of research and science to put Team USA athletes on the podium at international competitions and especially the Olympic Games. The group sponsors high value technology projects in many different Olympic sports with the goal to improving performance and preventing injury.

The diving project was one of these funded projects. The project was to develop a sensor that would mount under the end of the springboard. It was to measure the angle of the board at takeoff, thus giving feedback to the diver as to how well the springboard was flexed, and hence how well the diver was performing the takeoff. A prototype was built and tested, and an article was written for the Microsoft MSDN magazine. An interesting turnabout occurred in that in order to validate the angles that the sensor was measuring, we used video. That led us to the realization that in this case we wouldn't need the sensor. We could make the needed measurements from video alone. Excitingly we had another project underway that was using AI and a single video camera to generate a stick figure of a diver's takeoff. It needed now sensors or markers on the athlete. It uses pose estimation alone to track the diver's motion. The ability to measure diving board flexion will be an addition to this system.

My position as sport technologist evolved into director of sport technology and innovation, and now I am responsible to help develop technology for all Olympic sports, and automatic motion capture is a big part of that effort. Pose estimation and machine learning are now becoming the next frontier for skill analysis in many sports. As I already discussed, there have been many technologies to measure motion in the past, but they have all been complicated, and expensive, plus they have required items to be attached to the athlete; markers, sensors, wires etc. Now with video-based AI, ML and pose estimation we will be able to analyze an athlete's motion without putting sensors on them or even touching them. This has been my dream for many years. This feature is very desirable because it fits into an athlete's training patterns without disturbing their flow. That is not to say that we won't use sensors anymore, we certainly will, but we will use them judiciously and keep their number to a minimum.

In recent times the standard paradigm of capturing, storing, calculating and displaying data all on one laptop has changed. Now it is more typical that the data is captured and uploaded to the cloud. This is appropriate for two reasons, one, it is more convenient to store and distribute the data to all who wish to review it, and two, more computing power can be called up as needed. This is especially relevant when using pose estimation and neural networks and when converting the data from 2D to 3D.

This book shows how to implement many of the technologies that I have discussed an in fact developed during my career. It is a tinkerers book. It is for anyone interested in applications of AI and data science for sports, health and fitness, and analysis of human motion. If you are an experienced data or sport scientist or a hobbyist, looking to understand AI better, this book should give you plenty of inspiration and practical examples that take you on the journey from the foundation of sports mechanics to machine learning models and experiments. For a sport practitioner, familiar with biomechanics and kinesiology, this book explains how to use new machine learning tools that can take your research to the next level. For a data scientist, this book shows applications and real-life models in computer vision, sensors, and human motion analysis. Each chapter comes with a notebook with code samples in Python.

The first chapters give a Machine Learning 101 for anyone interested in applying AI in sports. Chapter 2, Physics of Sports, takes you through the foundations of biomechanics that you can find helpful to create and train machine learning models: starting with mechanics, kinetics, laws of motion and inertia, kinematics, and machine learning applications. Examples include using neural networks to predict a projectile range and calculate figure-skater's rate of spin. In Chapter 3 we will go over a sports data scientist toolbox: data science tools you'll need to work with AI and machine learning models, including Python, computer vision, machine learning frameworks and libraries. In Chapter 4, the book covers principles of neural networks, neurons, activation functions, and the basic single and multi-layer networks of neurons, as well as how to train a neural network. In Chapter 5 the book discusses sport sensors and devices you can use as a sport data scientist: Sensors (Deep Vision – Edge devices – Inertial movement sensors IMUs – Attitude and heading reference systems AHRS – Inertial and navigation systems GNSS – Range Imaging Sensors LIDAR – Pressure sensors – EMG sensors – Heart rate sensors). In Chapters 6-8 the book covers computer vision and AI methods for sports, as well as training machine learning models that can recreate human body motion in 2D and 3D. In Chapter 9 the book covers video action recognition: one of the most sought-after tasks for coaches and training, classifying movements and movement analysis. In Chapters 10-12 the book discusses advanced tasks in AI, such as reinforcement learning and ways to extend your research to modern cloud-based technologies.

In conclusion I am very pleased that you have chosen Kevin's book to improve your understanding of how AI can be applied to health, fitness and sports in a very practical manner. Enjoy the book and have fun experimenting as I did when I was a member of my high school radio club! Technology may have changed but curiosity certainly has not.

USOPC Disclaimer

The views and information in this material are my own, not that of the USOPC or any of its members or affiliates. The material may not reflect their views or positions.

Introduction

To my wife Katya and my 'Ohana

I was closing my winter season as a ski instructor in Breckenridge, Colorado, when I applied for a job at Microsoft. It was the time that they call the "mud season" in the mountains, when the snow starts melting everywhere, and there's not really much for a ski instructor to do until the next winter. I was very happy, when the offer came through during the summer, but I felt a little sad parting with my beloved Rocky Mountains and a ski town, as my job required me to move to sunny California, to the heart of Silicon Valley. So began my journey with one of the most amazing companies in the world, with the brightest people and great projects. I soon fell in love with California, its majestic sequoias, ocean waves, mountains, and golden hills, and true to my passion for sports and skiing, I kept contributing to the world of sports, health, fitness, and athletics. This book and my sport apps, Active Fitness and Winter Sports, research and ideas that came with them, are all part of this contribution. I feel lucky working with Olympic athletes, US Olympic Team, Special Olympics, US Ski Team, Professional Ski and Snowboard Instructors of America (PSIA-AASI), WTA, WNBA, my fellow professional ski and snowboard instructors, skiers, snowboarders, surfers, track and field athletes, tennis players, gymnasts, companies involved in manufacturing of sport equipment, and anybody making first steps in sports.

As an engineer working with the bleeding-edge tools and technologies, I was often asked for guidance from coaches, athletes, and sport equipment manufacturers: "How can AI help?" And I hope that this book can provide some ideas and practical methods. It is also meant to inspire those who learn about AI: data science is fun with sensors and sports! Today, data science is no longer a job confined to the office: a sport scientist is out there, with athletes and coaches, on the ski slopes, surfing the waves, discovering new ways to apply technology in real-life scenarios of sports, health, and fitness.

In addition to materials supplied with this book, you can also find a wealth of information about my research, at my website: http://activefitness.ai.

Kevin Ashley
California, May 2020

Getting Started

The noblest pleasure is the joy of understanding.

—Leonardo da Vinci

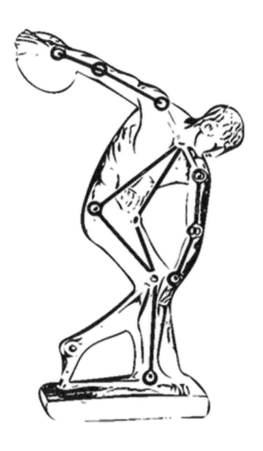

Machine Learning in Sports 101

I have always been convinced that the only way to get artificial intelligence to work is to do the computation in a way similar to the human brain.

—Geoffrey Hinton, "Godfather" of Deep Learning

© Kevin Ashley 2020

K. Ashley, *Applied Machine Learning for Health and Fitness*,
https://doi.org/10.1007/978-1-4842-5772-2_1

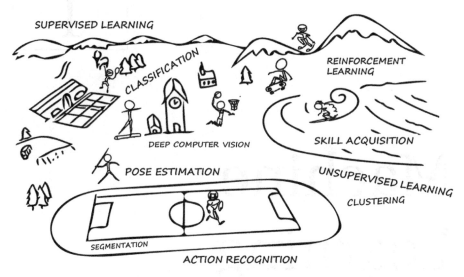

SUPERVISED LEARNING

CLASSIFICATION

REINFORCEMENT LEARNING

DEEP COMPUTER VISION

SKILL ACQUISITION

POSE ESTIMATION

UNSUPERVISED LEARNING

CLUSTERING

SEGMENTATION

ACTION RECOGNITION

Figure 1-1. Machine learning in sports

Getting Started

I don't know anything about luck, but that the harder I train, the luckier I get.

—Ingemar Stenmark, World Cup Alpine Ski Racer

In sports and athletics, results are achieved through *training* and *repetition*: machine learning is very similar. To train a skill, a human athlete needs thousands of repetitions. Training a movement skill for a humanoid robot, using machine learning methods like reinforcement learning (RL), requires tens of thousands or millions of iterations.

Machine learning is a relatively new method in sport science, but it's making huge advancements and already impacts many areas of sports, from personal training to professional competitions. For example, the International Federation of Gymnastics (Figure 1-2) announced that an AI judging system is about to be introduced to the world of professional competitions about the same year this book gets published! The system built for judging gymnastics is based on computer vision, sensors, and many of the same machine learning principles and research you'll discover from reading this book.

Figure 1-2. Gymnastics AI helps judging world-level competitions

For a coach, movement analysis is key to improving athletic performance and preventing injuries. In plain words, a coach can tell you how to become better at sports and not hurt yourself. Sport scientists are familiar with kinesiology and biomechanics and applying principles found in dynamics and classical mechanics for movement analysis. So, why machine learning?

I hope that this book helps answering this question, with practical examples a sport scientist or a coach can use. In addition to materials supplied with the book, check out ActiveFitness.AI (http://activefitness.ai) for additional materials, including videos, links to supplemental code, research, blogs, and apps.

Areas of Machine Learning

There're several areas or paradigms in machine learning that define most of the methods we'll be dealing with in this book: *supervised*, *unsupervised*, and *reinforcement learning*. This classification is open; in fact if you dig deeply into machine learning research and theory, you'll also discover *weakly supervised*, *self-learning* and a wealth of other methods. In this book, you will find practical projects and applications of these main areas of machine learning in health, fitness, and sports.

Supervised learning deals with datasets that include labeled data. Typical tasks for supervised learning include classification, for example, classifying activities or objects on the image. For supervised learning to work, large labeled datasets are required with input labels to train models (see Figure 1-3). Fortunately, you don't need to do most of image classification from scratch, datasets such as ImageNet contain tens of millions labeled images, and with techniques like transfer learning, you could reuse them in your model.

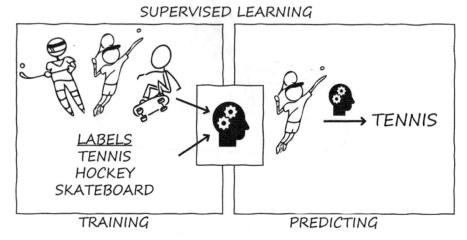

Figure 1-3. Supervised learning

Unsupervised learning doesn't assume that data is labeled; instead, its goal is finding similarities in the data, like grouping similar activities (Figure 1-4). It's often used for self-organizing dimensionality reduction and clustering, such as K-means. For example, if you train an unsupervised model with sufficient data containing images of athletes performing actions in different sports, such a model should be able to predict what group or sport a given image belongs to. This method is great if you don't have a labeled dataset, but sometimes you have *some* labels in an unlabeled set: this scenario is often called a *semisupervised* problem.

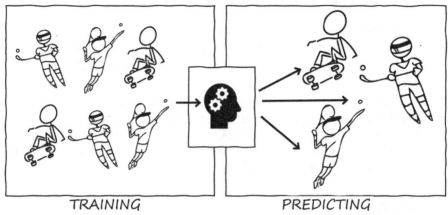

Figure 1-4. Unsupervised learning

Reinforcement learning (RL) applies a concept of an "agent" trying to achieve a goal and receiving rewards for positive and penalties for negative actions (see Figure 1-5). It originated from game theory, theory of control, and Markov decision process: it is widely used for robot training, including autonomous vehicles. This book goes over several applications of reinforcement learning in sports: for movement analysis, simulation, and coaching, check Chapter 10, "Reinforcement Learning in Sports," for more.

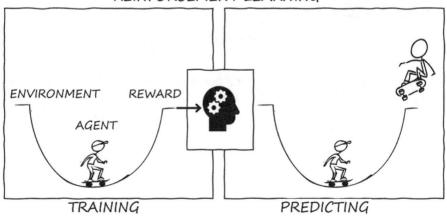

Figure 1-5. Reinforcement learning

Logic and Machine Learning

We just think you can have these great big neural nets that learn, and so, instead of programming, you are just going to get them to learn everything.

—Geoffrey Hinton

So, what in machine learning is different from a coder programming an algorithm? It's easier to illustrate if you consider what makes an algorithm work: usually it's logical rules that define how the algorithm handles the input data. With classic methods, we start with known rules and apply them to collected data to get the answers (Figure 1-6).

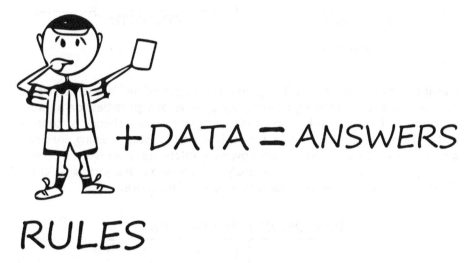

Figure 1-6. In classical algorithmic programming, we begin with rules and data to get answers

In machine learning we begin by giving our AI answers and data; the result is an AI "model" that contains rules that AI learned by observing inputs (Figure 1-7).

Figure 1-7. Machine learning works by training a model with answers and data; the result is a trained model that contains inferred rules

This looks simple enough, but it represents an entire paradigm shift in our approach to learning and computer programming. If you think about it, machine learning sounds almost too easy! In fact, most machine learning frameworks include only two essential methods: train and predict. The model needs to be trained, and then it predicts the outcomes.

Without getting deep into computer science, machine learning is certainly a powerful way to solve complex problems, but it is not a panacea for every task: understanding where it can help in health and fitness is part of the reasons for writing this book. Oftentimes, basic statistical methods, such as finding correlations between data points, regression, and classification, as well as algorithmic methods can be used, before bringing AI: machine learning deals specifically with training predictive models. Most machine learning models today are also not great at generalizing. It means that while they are trained on a specific set of data, prediction accuracy may drop significantly as you expand the inputs. With these words of caution, you'll see that these models just work in many applications, including sports!

Projects and Code

Figure 1-8. For chapters with practical code samples, check source code in the note-books supplied with this book and the video course with detailed walkthroughs at my Web site http://activefitness.ai

Projects and Code

Most chapters in this book include practical projects (Figure 1-8), code examples, and tips that you can use as a sport scientist. All sample code is in notebooks accompanying practical examples in the chapters, so locating source code is very easy: simply open a notebook corresponding to the number of the chapter you are reading, and you can follow through with the code. Notebooks are very popular with data scientists, and Python is one of the most widely used languages in data science, so it was natural to use it for most examples in this book.

■ **Code tips**   Check notebooks, corresponding to the chapters you are reading for code samples and practical projects.

Introducing Tools

I already mentioned notebooks and Python being the language of choice for many data scientists. In this book you'll go over many practical projects that use various classic machine learning frameworks, such as scikit-learn, PyTorch, Keras, and TensorFlow, as well as some specialized frameworks and libraries such as OpenAI Gym. The idea is to show a sport data scientist a broad range of tools. Note that for data science, you often need a hardware-accelerated system, as most of the tools can take advantage and work much faster with graphics-accelerated processors (Figure 1-9).

Figure 1-9. Learning new methods often requires learning new tools; for machine learning, it also means appropriate hardware!

If you are just beginning with computer science, or if you are an expert, you'll find plenty of useful practical projects in this book using a variety of modern tools. In addition to libraries and tools, a data scientist often needs large datasets to train models. After playing with your data science project, you may need to scale your models and deliver them to users. Making your AI available for apps, bots, and other services shouldn't be an afterthought. This book also provides chapters that cover working with the cloud and consuming your models. Machine learning may be computation intensive, so getting a graphics processing unit (GPU)-enabled device is a plus: you can also use cloud-based services from Microsoft Azure, Google, Amazon, and others (Figure 1-10).

■ **Tools tips** For in-depth dive into modern data science tools I used in this book, as well as how to get started on most code samples, I recommend reading Chapter 3, "Data Scientist's Toolbox."

Figure 1-10. Data science can be computation intensive: you'll need a capable machine with a GPU to train the model or cloud resources to scale

Neural Networks

All you need is lots and lots of data and lots of information about what the right answer is, and you'll be able to train a big neural net to do what you want.

—Geoffrey Hinton

Neural networks appear magical to many people, including data scientists themselves! Neural networks are mathematical abstractions of what we think is the learning process in our brain, although that is not an accurate approximation, for several reasons. For one, we still don't know exactly how human brain works and learns. And although neural networks as mathematical models are similar in some way to the way neurons work, some learning mechanisms, like backpropagation, although present in convolutional neural nets (CNNs), have not been discovered in biological neurons.

The first attempts to create a computational model of a neuron began in the 1940s, with the first perceptron machine built almost 70 years ago, but it's not until data science introduced multilayer convolutional neural nets with learning mechanisms such as backpropagation that they became truly successful.

In this book we'll use many practical applications and projects using neural networks: for classification, semantic segmentation, video action recognition, and deep vision. Check Chapter 4, "Neural Networks," which covers neural nets in depth with several practical examples.

Deep Vision

ImageNet is a dataset of over 15 million labeled high-resolution images belonging to roughly 22,000 categories.

—Alex Krizhevsky

Deep vision is an area of computer vision that uses deep learning methods. The earliest information we recorded from the ancient times was visual, from petroglyphs depicting action scenes dated back almost 30,000 years to ancient Greek art of the first Olympic Games, first recorded motion pictures, and finally YouTube with billions of videos. As data scientists, taking advantage of the abundance of picture and video data is a natural step toward understanding movement and motion in sports. Typically tasks we want to do with deep vision are:

- **Classification** of objects is essential for a machine learning model to understand various classes of objects present in the video or image: for example, classifying a human or a tennis ball is the first practical step toward analyzing the image. Most image classification models are trained on ImageNet and use CNNs.

- **Detection** of objects from images or videos (Figure 1-11) is a fundamental task in deep vision: typically, it involves finding a bounding box for an object. Detection is usually done with R-CNN (Region-based CNN).

- **Semantic segmentation** is the next step from classification and detection, dividing the image to segments on the pixel level. Because objects are often occluded in images, with parts of one blocking another, *instance segmentation* helps identifying entire objects regardless of occlusion. Some examples of networks that semantic segmentation models use are Mask R-CNN.

- **Object tracking** is another task in deep vision that deals with video and objects displacement over time.

Figure 1-11. Applying deep vision methods, such as image detection and classification, allows us to find a surfer and a surfboard on this image (see Part II for practical code examples)

These deep vision tasks provide a foundation for analyzing complex objects, such as human body in motion. From biomechanics, we know that a physical model for a human body can be represented as a set of connected joints. Can deep vision be used to infer human body key points from a picture or video? To answer this question and demonstrate practical applications, in the chapters dedicated to computer vision, we'll cover these methods of applying deep vision to human body analysis:

- **2D pose estimation** is the task of estimating human body pose, typically based on the model that consists of human body parts and joints (Figure 1-12).

- **3D pose estimation** is the next step from 2D pose estimation that attempts to reconstruct a 3D environment and the human pose in that environment.

- **Action recognition** is another area that's typical for sports and movement analysis. Recognizing actions from basic walking, sitting, to deep action recognition such as "a skier performing a carving medium radius turn" is covered in Part II.

Figure 1-12. A more advanced model, detecting joints of the human body (for examples on body pose estimation, see Part II)

In this book I'll show you practical projects for detection, classification, semantic segmentation, object tracking, and body pose estimation, from still images to video. All these methods can help any sport or data science practitioner, because image and video data is so abundant. We'll be using the same generalized machine learning frameworks like scikit-learn, PyTorch, and Keras and computer vision libraries such as OpenCV. In chapters of this book dedicated to deep vision, you'll learn about tools; datasets, such as Common Objects in Context (COCO), Sports-1M, and Kinetics; models; and frameworks, as well as many practical projects for pose estimation, video action recognition, and applications in sports.

Sensors

From reading the first pages of this book, you probably figured that machine learning is dependent on data. Part of the reasons sport-related research in machine learning has been so successful is because of huge amounts of data we collected in sports. We can use video and photo with computer vision as the ground truth for machine learning models, but that approach comes with a caveat: video typically includes occlusions, where parts of the picture are hidden from the camera. That means, that if we train a model based on video data, that model will not be as accurate as the one that uses sensor data for ground truth measurements (Figure 1-13).

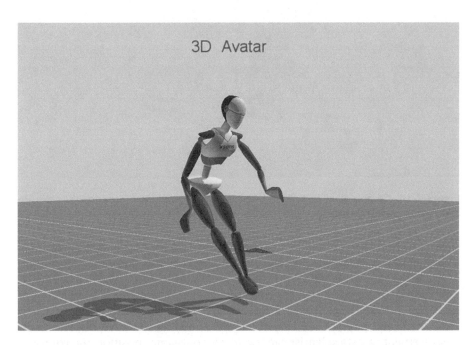

Figure 1-13. A skier performing slalom turns, wearing a high-quality sensor suit from Xsens integrating 17 sensors allowing full-body mocap at 240Hz; this data can be used to provide ground truth for model training

This is where biomechanics and sports intersect with areas such as robotics, motion capture (mocap), and gaming: it turns out that high-quality mocap data is also needed for a gaming character visualization, as well as training robots. In this book, you'll find practical examples for 3D and 2D pose estimation, training models to simulate and perform humanlike actions, and other examples that can be applied in sports as well as gaming and robotics.

Figure 1-14. Sensors allow capturing motion at a very high accuracy, providing ground truth data for the model training

The most basic sensor used to capture motion is an inertial movement unit (IMU) usually containing an accelerometer, a gyroscope, and a magnetometer. These sensors provide data readings on acceleration and rotation at a rate of 100–200Hz or samples per second, a frequency that's much higher than a recorded video which usually gives 30–60 frames per second. With some basic math and physics, it's easy to derive speed and relative displacement, measurements a data scientist can use to compute joint position and rotation. High-quality full-body sensor suits such as Xsens provide 17 or more sensors placed against the joints to provide a full 3D body motion capture with a very high accuracy for fast action sports (Figure 1-14). Check Chapter 5, "Sensors," for an in-depth dive into various types of sensors used in sports that can provide data for machine learning models.

Reinforcement Learning

You can go right or left, but you can't very well do both at once.

—The Endless Summer

Reinforcement learning (RL) is a method of machine learning that resembles the way we learn by trying and finding the best way to perform actions. This machine learning method does not require a supervised dataset or a labeled set of data to train. In reinforcement learning an agent explores the environment by performing actions and receiving rewards for actions that lead to achieving goals (Figure 1-15).

This sounds a lot like a coaching problem. Check Chapter 10, "Reinforcement Learning in Sports" in this book, for a deep dive into practical projects that use RL methods in sports, including skateboarding, surfing, snowboarding, and gymnastics. You'll also learn some of the most popular tools like OpenAI Gym and RL Baselines and physics libraries such as PyBullet and RL model zoos and will be introduced to various reinforcement learning environments, methods such as actor-critic (A2C), Deep Q Networks (DQN), and others.

Figure 1-15. Some classic RL environments like MountainCar can be readily projected to sports, for example, a skateboarder riding half-pipe, or a surfer riding a wave

Reinforcement learning helps solving many simulation problems, for example, visualizing and training the model to execute athlete's movements exactly the way they are performed, mimicking, for example, a tennis serve of Roger Federer. Reinforcement learning typically is a computation-intensive method that requires many iterations to train, for example, it may require up to 60 million epochs to train a skill, that is, two days 8 core GPU, but given how quickly hardware evolves, in the future that shouldn't be a problem.

Summary

As Robert Redford put it, "sport is a wonderful metaphor for life," and modeling life experiences is part of data science. In this chapter I introduced concepts and areas of machine learning I'll be covering further in detail in this book. If you are a data scientist or a sport science practitioner interested in machine learning, you will find many practical examples that can help you in real life. From deep computer vision to working and applying sensors in the field of various sports, analyzing and processing data and training models, new technologies and projects collected in this book should help adding machine learning as a tool. With sports and human movement analysis, there're many intersections with many other industries, including gaming, robotics, and motion capture and simulation.

You can find additional materials related to the research I've done in this book, including videos and apps at my ActiveFitness. AI website: `http://activefitness.ai`.

Summary

Physics of Sports

No great discovery was ever made without a bold guess.

—Isaac Newton

Figure 2-1. Machine learning can predict gravity, but it may need a bigger training set than Sir Isaac Newton's apple

Overview

This chapter provides an overview of physical principles used in modern sport science. In addition to physics, kinesiology, and biomechanics, we will also discuss how deep learning can help a sport data scientist, and vice versa, how we can improve our models by knowing a few physics principles. Classical mechanics is a reliable method of movement analysis, and it's a valuable tool if you're planning to build any practical sport machine learning models. In this chapter, I'll show how machine learning models, including neural nets and reinforcement learning, can be applied to biomechanics.

The classical biomechanical approach to sport movement analysis implies laws of physics. Essentially, those are principles our science learned from the universe through centuries of research or smart guesses, like Newton's apple (Figure 2-1). How can machine learning and modern data science help? In machine learning we train our model to *learn* rules, based on the data that we provide as a training set. Using our knowledge of laws of physics also helps validating and training our models. This idea inspired, for example, research around PGNN (physics-guided neural net): the idea that a typical loss function can be combined with finding a physical inconsistency.

Can machine learning be combined with biomechanics to provide that "shortcut" inferred from centuries of learning laws of physics? Absolutely! In this chapter I'll show some practical examples on how this can be achieved. Just like in data science, when transfer learning is used, our model would benefit from knowledge transfer from physics. In fact, there's already research that uses neural networks to discover physical laws and compress them into simple representation, such as differential equations. There're many opportunities in combining our knowledge of physics with machine learning; let's go over the fundamental principles of physics used in sport science.

Mechanics

Motion in sports can be described by mechanical models. A course of kinesiology includes biomechanics, science that applies mechanical models and laws to analyze human body movement. Mechanics involves studying motion of rigid bodies, solids, and deformable bodies and fluids. Biomechanics is primarily concerned with physics of human body and movement and includes the following areas:

- **Kinematics** describes effects of linear and angular motion: velocity, displacement, acceleration, and position.

- **Kinetics** explains what causes motion: forces, torque, and moments.

- **Work, energy, and power** are areas where physics helps sport science determine efficiency, calories spend, and fatigue.

- **Skeletal, joint, and muscular mechanics** of body tissues, joints, and the human frame studies interaction between body parts, stress and strain (this area immediate application to body pose estimation in machine learning).

An experienced sport scientist can use biomechanics to provide movement analysis for any sport. Any coach can use these models because we all know physics from school.

Kinetics: Explaining Motion

In this section I'll show you some applications of linear and angular kinetics of rigid bodies in sports. To make it interactive and fun, the accompanying Jupyter Notebook contains Python code you can follow to run and visualize examples.

First Law of Motion (Law of Inertia)

The first law of motion says that if a body is at rest or moves at a constant speed, it will remain at rest or keep moving at a constant speed in a straight line, unless it is acted upon by a force.

In physics, the concept of momentum was introduced before Newton by Descartes, meaning the amount of motion. The definition of momentum is the product of the mass of the body to its velocity. The unit of momentum in the International System of Units (SI) is kg*m/s:

$$momentum = mass * velocity$$

$$p = mv$$

Figure 2-2. How much force does it take to stop a running football player?

In football (Figure 2-2), a collision occurs when a tackler stops a running quarterback. Let's say the mass of a quarterback is 100 kg and his speed 5 m/s. Then the momentum of the quarterback from the first law of motion is 500 kg*m/s:

■ **Code examples** You can find all source code for the book in accompanying Jupyter notebooks, corresponding to each chapter. Unless stated otherwise, the source code is in Python. For a deep dive into data science tools and configuring your data scientist toolbox, see Chapter 3, "Data Scientist's Toolbox."

```
m = 100. # kg
v = 5. # m/s

# Momentum of a football runner
p = m * v
print(f'Momentum {p} kg*m/s')
```

The impulse it takes for a tackler to stop a quarterback is defined by the speed and time of collision. Let's say the tackler occurs to full stop, when the speed is zero, and the time of collision is half a second, then the impulse:

```
v1 = 0. # m/s
dt = 0.5 # s

# F=ma or F=m*dv/dt
# F*dt = m*dv

# Impulse to stop a quarterback
dv = v1-v
I = m * dv * dt
print(f'Impulse to stop {I} N*s')
```

Second Law of Motion

Newton's second law relates acceleration of a body with an external force F acting on it:

$$force = mass * acceleration$$

$$F = ma$$

The unit of force is newton (N) = kg*m/s^2. In our earlier example, the force with which the tackler acts on the quarterback is:

```
# Force exerted by tackler on the quarterback
F = m * dv / dt
print(f'Force {F} N')
```

Note that in this example, acceleration is implied by using the time derivative of velocity. Since acceleration is the rate of change of velocity, the second law also means that force is the rate of change of momentum.

Third Law of Motion

For every action, there's an equal and opposite reaction. For example, when a player hits the volleyball (Figure 2-3), the ball reacts with an equal force on the player.

Figure 2-3. A volleyball player hits the ball, which reacts with an equal force on the player

The laws of motion introduced earlier can also be expressed for rotation, or angular movement. Many sports use angular motion: spinning, swinging movements. Same mechanical laws apply for angular motion, but instead of forces, we use toque, linear displacement becomes angular displacement, and so on.

Kinematics: Projectile Motion

Sports often involve projectiles: for example, javelin, balls, jumpers, and so on. Projectile is any object that is released with initial velocity and is subject to forces of gravity and air resistance.

In sports, javelin (Figure 2-4) and discus were included in Olympic sports since the original ancient Olympic Games, dating back to 776 BC. The flight of both javelin and discus is an application of both kinematics, because it can be described by projectile motion and aerodynamics.

Figure 2-4. Javelin as a sport of throwing a spear for distance, included in the ancient Greek Olympic Games

The factors of kinematics for projectile are release speed, release angle, and release height. The aerodynamics effect comes from the lift during the flight and air resistance. The aerodynamics factors are angle of attack and wind speed (Figure 2-5).

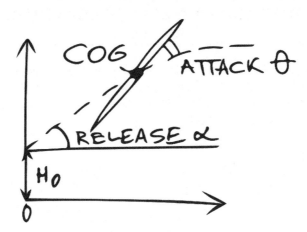

Figure 2-5. Javelin flight: angle of release and angle of attack relative to center of gravity (COG)

In this example I'll focus on kinematics aspects of the motion; my goal is to show how physics can be combined with machine learning effectively and used in practical sport applications. A projectile displacement along horizontal and vertical axes can be determined as:

$$x = v_0 t \cos \alpha$$

$$y = v_0 t \sin \alpha + \frac{1}{2} g t^2$$

Given the release speed v, release angle α, and release height h, the range of a projectile can be calculated using the formula:

$$x = \frac{v^2}{2g} \sin 2\alpha \left(1 + \sqrt{1 + \frac{2gh}{v^2 \sin \alpha}} \right)$$

or, when the projectile is launched at the ground (i.e., h=0):

$$x = \frac{v^2}{2g} \sin 2\alpha$$

Formula 2-1. Projectile range when launched from the ground

PROJECT 2-1: CALCULATE AND PLOT A PROJECTILE TRAJECTORY AND FIND BEST RANGE AND RELEASE ANGLE

In this project we'll find an optimal angle of release for an athlete throwing a projectile. Let's assume the release speed of a projectile at 30 m/s and for simplicity it's released at the ground. To test various angles of release, we'll create a list of test angles from 20 to 50 degrees spaced at 5 degrees:

```
import numpy as np
import matplotlib.pyplot as plt
%matplotlib inline

g = 9.81 # m/s^2
v = 30 # release speed m/s
angles = np.arange(20, 50, 5)
```

Finally, to plot the results and iterate through test angles finding the best angle (Figure 2-6), determined by maximum horizontal displacement:

```
max_distance = 0
best_angle = None
t = np.linspace(0, 5, 300)
for angle in angles:
    x,y=projectile(angle*np.pi/180, t)
    distance = x[-1]
    if max_distance < distance:
        max_distance = distance
        best_angle = angle
    plt.plot(x, y, label=r"$\alpha$"+f"={angle:.0f} d={distance:.1f}m")
    plt.legend(bbox_to_anchor=(1, 1))

plt.show()
print(f"max distance: {max_distance:.1f}m angle={best_angle:.1f}")

Output: max distance: 91.5m angle=45.0
```

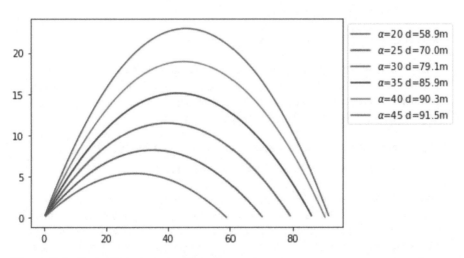

Figure 2-6. Projectile trajectories for various angles

PROJECT 2-2: TRAIN A NEURAL NETWORK TO PREDICT A PROJECTILE RANGE

We can also train a machine learning model to predict projectile range. In machine learning, finding a continuous value is a problem of regression. We will use Keras to build the model, and scikit-learn to generate a dataset for training. Since we know the formula for the range of projectile (see Formula 2-1), generating a training dataset should be easy:

```python
import numpy as np
from sklearn.model_selection import train_test_split

g = 9.81

def generate_data(size=1000):
    v = np.random.uniform(5, 35, size)
    alpha = np.random.uniform(20,60, size)
    projectile_range = [np.power(v,2)*np.sin(np.deg2rad(2*alpha)) / g]
    y = np.reshape(projectile_range, (size, 1))
    return train_test_split(np.vstack([v,alpha]).transpose(), y, test_size=0.2, random_state=42)

X_train, X_test, Y_train, Y_test = generate_data()
```

I set the dataset size to 1000 by default, and release speed is in the range 5–35 m/s and release angle 20–60 degrees. Resulting data is automatically split into training and test set, the input X is an array with rows containing release speed and release angle values, and the output contains predicted range:

```
Input set: (800, 2) sample value: [21.41574047 42.88498157]
Predicted set: (800, 1) sample value: [46.62432173]
```

Next step is building a network in Keras to predict the projectile range. I used a simple network with two hidden layers, and the last layer is a linear layer with no activation. The mean squared error (MSE) loss function is used a lot in regression models. The MAE (mean absolute error) metric is the value difference between the predictions and the targets. For example, a MAE of 0.99 means that you are off by 99 centimeters from the target range:

```python
from keras import models
from keras import layers
from keras import callbacks
```

```
def build_model():
    model = models.Sequential()
    model.add(layers.Dense(64, activation="relu", input_dim=2))
    model.add(layers.Dense(64, activation="relu"))
    model.add(Dense(1))
    model.compile(optimizer='rmsprop', loss="mse", metrics=['mae'])
    return model
```

Now the network is build; let's train it! The best model will be saved in projectile.hdf5 file.

```
model = build_model()
checkpoint = callbacks.ModelCheckpoint(filepath="projectile.hdf5", ver-
bose=1, save_best_only=True)
model.
fit(X_train, Y_train, epochs=500, batch_size=10, callbacks=[check_point])
score = model.evaluate(X_test, Y_test)
Y_pred = model.predict(X_test)
print('Score:', score)
```

The best score is within a meter from the projectile formula; just to make sure, let's check results from the model with the actual formula (your results may vary as dataset generation for model training is random):

```
angle = 45
speed = 30
prediction = model.predict(np.array([(speed,angle)]))
actual = speed**2*np.sin(np.deg2rad(2*angle)) / g
print(f"prediction: {prediction} actual: {actual}")
```

Output

```
prediction: [[91.703804]] actual: 91.74311926605505
```

Not bad, our model predicted projectile landing within 4 cm of the target, and that's with a thousand training samples. Of course it's not one apple that the story tells us prompted Newton to discover laws of motion, but our neural net was able to learn the hypothesis of projectile flight pretty quickly!

Angular Motion

Angular movement is common in sports and can be described using the same dynamics principles we discussed earlier.

Angular First Law (Law of Inertia)

Similarly to the first law for linear motion, the law of inertia can be applied to rotational systems, but instead of linear velocity, we refer to angular velocity, and force is torque in rotational terms. In angular motion, momentum relates to the angular momentum:

angular momentum = moment of inertia ∗ angular velocity

$$L = I\omega$$

Every object moves with a constant angular velocity unless acted upon by a torque.

Angular Second Law

Newton's second law of rotation is called the law of rotational dynamics. It states that angular acceleration is proportional to net torque and inversely proportional to moment of inertia:

torque = moment of inertia ∗ angular acceleration

$$\tau = I\alpha$$

Angular Third Law

For every applied torque, there's an equal and opposite torque. For example, for a figure skater, gravitational force creates torque toward the ice, but that also means that the ice provides an opposite torque pointing in the opposite direction.

Conservation Laws

In physics momentum conservation laws state that *in a closed system momentum doesn't change over time.* This applies to both linear and angular momentum discussed earlier. In sports conservation laws are used in many types of activities, for example, in figure skating.

A spinning figure skater (Figure 2-7) mechanically is a nearly closed system; we will make this assumption for the purpose of this example. The contact that the skater maintains with the ice has friction that is negligibly small and rotation happens around the pivot point. Her angular momentum is conserved, because there's no torque acting on her. The rotational first law (law of inertia) states that angular momentum remains constant:

$$MOI_{out}\omega_{out} = MOI_{in}\omega_{in}$$

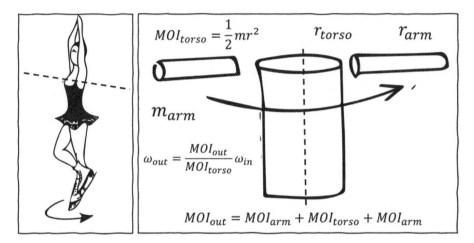

Figure 2-7. Figure skater and conservation of angular momentum

To accelerate, the skater can pull her arms, the moment of inertia *MOIin* will decrease, and angular velocity ω, therefore, will increase. When she's extending her arms, the moment of inertia *MOIout* increases; this results in decrease of ω so she slows down. How much does she accelerate when she pulls her arms out? Let's estimate her moment of inertia with arms in and out, and also increase in angular velocity. Assuming the skater weighs 55 kg and making some assumptions about body parts weight and proportions (radius) of her torso, and length of arms, it's easy to calculate momentum. I used cylinder as an approximation for the athlete's body:

```
import numpy as np
import matplotlib.pyplot as plt
%matplotlib inline

g = 9.81 # m/s^2
```

```python
m_body = 55 # kg
m_torso = 0.5 * m_body # kg
m_arm = 0.06 * m_body # kg
print(f"m_arm: {m_arm:.2f}kg\nm_torso: {m_torso:.1f}kg")


r_torso = 0.25 # m
r_arm = 0.7 # m

MOI_torso = (1./2.) * m_torso * r_torso**2
MOI_arm = (1./3.) * m_arm * r_arm**2
print(f"Moment of inertia (arm): {MOI_arm:.3f}kg*m^2")
print(f"Moment of inertia (torso): {MOI_torso:.3f}kg*m^2")


MOI_1 = MOI_arm*2 + MOI_torso
MOI_2 = MOI_torso
print(f"Moment of inertia (out): {MOI_1:.3f}kg*m^2")
print(f"Moment of inertia (in): {MOI_2:.3f}kg*m^2")


w1 = 2 # revolutions per second

w2 = w1 * MOI_1 / MOI_2

print(f"Spin rate In: {w1:.0f} Out:  {w2:.0f} rev/sec")

Output:
m_arm: 3.30kg
m_torso: 27.5kg
Moment of inertia (arm): 0.539kg*m^2
Moment of inertia (torso): 0.859kg*m^2
Moment of inertia (out): 1.937kg*m^2
Moment of inertia (in): 0.859kg*m^2
Spin rate In: 2 Out:  5 rev/sec
```

The moment of inertia of the figure skater increases and her spin rate may increase from 2 to 5 revolutions per second. Sounds like our skater can spin for a long time, but why does she need to accelerate? Well, in the beginning, our assumption was the absence of external torque, but in reality, there's at least the force of gravity acting on the skater and there's also skates friction. That gravitational force wants to pull the skater down (Figure 2-8), and the torque will make the skater slowly revolve around z axis. This is called precession, and it has precession velocity ωp. In order to stay in balance, the figure skater needs to accelerate.

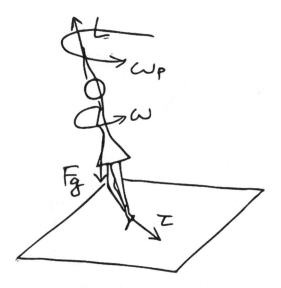

Figure 2-8. Figure skater precession movement

Energy, Work, and Power

The ability to do work is called energy, another key physics concept we use in sports. Biomechanics is using several forms of energy that mechanics defines: kinetic energy as the energy resulting from motion, potential energy due to position as a result of gravitational force (e.g., a skier stranding on the top of the run; Figure 2-9), or strain energy (a form of potential energy stored due to elastic properties of materials). Any moving object has kinetic energy due to its motion:

$$K = \frac{mv^2}{2}$$

Figure 2-9. Potential energy of skier and surfer stored, respectively, in the elevation of the run and the height of the wave

Potential energy, due to gravitational force, is defined by acceleration due to gravity g (9.81 m/s^2) and h (height):

$$P = mgh$$

Elastic potential energy is due to deformation and can easily be explained by looking at a tennis racquet: the tension of the strings impacts the serve. The stiffness is defined by constant k, and the amount of deformation is Δx:

$$E = \frac{k \Delta x^2}{2}$$

Now, we can also introduce the physical concept of work:

$$work = force * distance$$

$$W = Fs$$

The unit of work is joules, and it's equal to energy transferred to an object when the force of one newton acts on it through a distance of one meter.

Physics and Deep Learning

Geometrical ideas correspond to more or less exact objects in nature and these last are undoubtedly the exclusive cause of the genesis of those ideas.

—Albert Einstein, Theory of Relativity

Reading the theory of relativity, it struck me how simply Einstein introduced concepts of relativity to the world that was taught classical mechanics based on ideas of "point," "straight line," and "plane," those fundamental concepts of Euclidean geometry. I'll try to use Einstein's argument to explain my point with classical mechanics and machine learning.

Classical mechanics employs concepts based on geometrical models: they are great at describing human body motion. In order to actually use these models, a sport scientist would typically use some data, whether visual, like a video or a picture, or sensor data recorded during a sport activity. Next, a sport scientist would draw a biomechanical model based on center of mass, force, torque, velocity, moment of inertia, and so on. Such an analysis also takes an experienced sport scientist with knowledge of physics to do. It is also very hard to do *automatically*: the process is largely analytical.

Machine learning uses observations (training data) as the input, and the output is generating the rules (weights of the model). Do you see where I'm getting to? In biomechanics we use existing rules, while our machine learning model may not necessarily be aware of the laws of physics, but it learns rules from observations! From training each epoch, our model learns what Newton, perhaps, learned by observing his apple falling. Our knowledge of physics comes very handy when we need to build and train models, for example, during validation (Figure 2-10).

Figure 2-10. The dream of sport data scientist: surfing with AI

Models

A physical model typically includes several *parameters*. How many? It depends on the model, but we typically deal with time, acceleration, mass, and so on. Let's say physical models typically have *several* parameters. A machine learning model has millions of parameters. To illustrate my point, with principal component analysis (PCA), a model we built to classify movements of a beginner skier compared with an advanced skier resulted in several thousand features that significantly contributed to the movement. When a human coach looks at a skier, he can do that classification immediately, but it's not because of biomechanical model he's using in his head! Just to give you a practical example, the video recognition model we use in Chapter 9, "Video Action Recognition," has 31 million parameters! In fact, it's easy to count parameters with this code snippet. I'm using PyTorch here, but similarly you can do with other frameworks (see Chapter 3, "Data Scientist's Toolbox"):

```
import torch
import torchvision
import torchvision.models as models

model = models.video.r2plus1d_18(pretrained=True)
model.eval()

params_total = sum(p.numel() for p in model.parameters() if p.requires_grad)
print(f'Parameters: {params_total}')
```

Output:
Parameters: 31505325

Of course, no human sport analyst would use so many parameters when creating a biomechanical model! The whole point of using physics here is that a physical model implies a lot of knowledge we have about movement and greatly simplifies understanding. AI has often been criticized as being a "black box," and rightfully so, $F = ma$ is a lot easier to understand than a NumPy array of weights in the model!

Interesting fact   Machine learning models, for example, for movement classification, work just like your brain, and that is without knowing biomechanics at all! But physics stores a lot of knowledge we can use in our models.

A coach knows how to quickly classify a beginner from an expert. A professional ski instructor is used to identify ten levels of skiers for a teaching progression, but it becomes harder as you get into the level of Lindsey Vonn and Mikaela Shiffrin to do movement analysis using the standard teaching scale. This is why Olympic-level coaches are hard to find! Machine learning is also great at what human brain is good at: classifying movements; given the right amount of data, it can do it at any level of sport action.

Mechanics and Reinforcement Learning

Reinforcement learning has been a hot area in deep learning research. As sport scientists, we cannot bypass frameworks with the name like OpenAI Gym with our attention: it sounds so close to applied field of sports that we should give it a shot. In reinforcement learning (RL), the idea that you can train the model by giving an agent a reward sounds like a great idea to apply to sports (Figure 2-11).

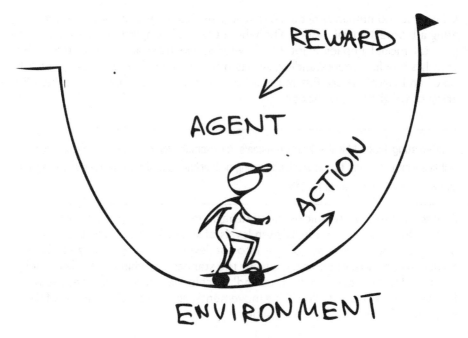

Figure 2-11. Reinforcement learning: agent acting in the environment and receiving a reward

In fact, reinforcement learning has many methods, including Q-learning, actor-critic (A2C), and so on. The fact that reinforcement learning can also solve continuous problems, like predicting velocity, displacement, and other measure, sounds especially interesting. Previously in this chapter, we used a neural net with linear regression to predict a projectile; now we'll use reinforcement learning for problems that often happen in sports like skateboarding, snowboarding, surfing, and skiing.

PROJECT 2-3: STARTING WITH REINFORCEMENT LEARNING FOR SKATEBOARDING

In sports like skateboarding (Figure 2-12), surfing, and skiing, there's often a need for an athlete to gain altitude, increasing gravitational potential energy for the next maneuver, and gain kinetic energy. Think of a half-pipe, for example: in half-pipe which is common in skateboarding, snowboarding, and skiing, athletes use physics of angular impulse and momentum we discussed earlier in this chapter to continuously move along the half-pipe and even perform tricks in the air outside of the half-pipe. The walls of half-pipe are relatively short, but the athlete can move along the half-pipe by continuously moving up and down the walls.

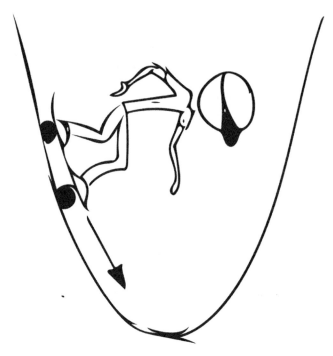

Figure 2-12. A skateboarder uses momentum to travel up the walls in the half-pipe

Similar idea is with surfing: one of the first skills surfers learn is moving up and down the face of the wave to move along the wave rather than straight down to shore. This provides room to countless surfing maneuvers, like cutbacks and air, and increases the distance surfers travel: in fact, according to Guinness World Records book, the longest surf ride was 10.6 miles (www.guinnessworldrecords.com/world-records/longest-surfing-ride-on-a-river-bore/)! The face of the wave was up to 8 ft. at the highest, yet because the surfer was moving up and down the face of the wave, he was able to have the longest ride!

Back to reinforcement learning, OpenAI Gym provides convenient environments specific to problems they are solving. You can always check the latest registry of available environments with:

```
import gym
print(gym.envs.registry.all())
```

For applications in skateboarding, snowboarding, surfing, and skiing, gym environments in classic control area are especially interesting; specifically, let's focus on MountainCarContinuous environment that solves a very similar problem! The following code creates the environment and creates a visualization of the problem:

```
from IPython import display
import matplotlib
import matplotlib.pyplot as plt
%matplotlib inline

env = gym.make('MountainCarContinuous-v0')
env.reset()
plt.figure(figsize=(9,9))
img = plt.imshow(env.render(mode='rgb_array')) # only call this once
for _ in range(20):
    img.set_data(env.render(mode='rgb_array')) # just update the data
    display.display(plt.gcf())
    display.clear_output(wait=True)
    action = env.action_space.sample()
    env.step(action)
env.close()
```

The MountainCarContinuous environment (Figure 2-13) provides an out-of-the-box model of a car that needs to use the momentum to get on top of the hill. That is very similar to what a skateboarder needs to do in order to skate to the top of the half-pipe wall. The difference is that the built-in environment includes the notion of power (i.e., the car is said to be underpowered, but still powered); obviously our skateboarder needs to use his body to create the momentum. In skateboarding, surfing, and other sports, this is often called "pumping"; by crouching down and then extending, the athlete is able to increase his velocity. For the purpose of our exercise, we can say that the ability to "pump" provides a mechanical equivalent of the engine. In the following chapters, I will show how to train a model that uses reinforcement learning to solve the problem of half-pipe.

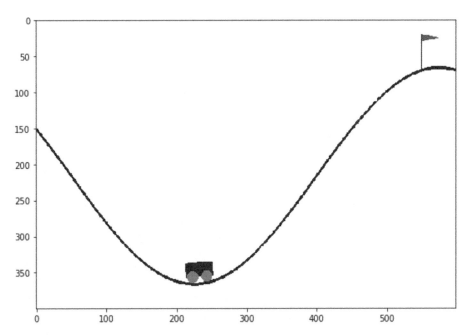

Figure 2-13. Reinforcement learning and skateboarding: using momentum to reach the top of the half-pipe

Summary

Physics principles describe motion and provide an analytical foundation for biomechanics, the science that studies human movement. In this chapter we had an overview of classical mechanics: kinetics and kinematics to analyze movement in sports, including linear and angular laws of motion and potential and kinematic energy. Practical projects included in accompanying notebooks for this chapter include code for examples in kinematics of projectile movement, a machine learning model that predicts a projectile using neural networks, mechanical analysis of figure skater spinning, and applications of reinforcement learning for skateboarding, snowboarding, surfing, and skiing.

Machine learning in practical sports is a relative newcomer; the way we teach sport science today will be changing a lot in the years to come, thanks to evolution of machine learning models, tools, and accuracy that's happening in data science and the field. In order for us to build models for sports motion and movement, understanding physics principles of motion is essential.

Data Scientist's Toolbox

Overview

We love our tools! Each job eventually comes up with a set of tools that define the job (Figure 3-1). In medieval Europe, for example, tradesmen and merchants formed guilds that defined trades and crafts for centuries. Coats of arms of these organizations typically included tools of the trade. If we were to form a guild of data scientists, our coat of arms would include many tools: languages like Python, libraries for deep learning and modeling like PyTorch or Keras, computer vision libraries like OpenCV, data processing libraries, tools for visualization, model deployment, cloud tools that operationalize our data science projects, and more (Figure 3-2). Studying human movement is a field combining kinesiology, biomechanics, data science, and oftentimes sensors, so in this chapter, we'll talk about our toolbox.

© Kevin Ashley 2020
K. Ashley, *Applied Machine Learning for Health and Fitness*,
https://doi.org/10.1007/978-1-4842-5772-2_3

Figure 3-1. Our tools define what we can do as data scientists

Figure 3-2. Imagine a coat of arms for data science

The list of tools available to data scientists is long and grows at an increasing pace, so in this chapter, I aim to introduce some of the most popular examples.

Languages

These days many data scientists use open source libraries and languages, such as Python, R, Lua, and Julia, that became very popular. Predating their popularity, MATLAB is one of the oldest commercial languages and tools for numerical computing, dating back to 1970. In this book, most of the code is written in Python, although some examples use libraries developed with lower-level languages, like C, which is commonly used in developing low-level or real-time code. The use of Python in data science grows faster compared to other languages, due to its approachability and flexibility. It's a popular, general-purpose computing language, so most data science packages are available in Python.

Data Science Tools

A typical workflow for a data scientist focused on sports, health, and fitness often involves working with coaches, trainers, and athletes, collecting data, and then analyzing it. In this book we treat each practical project as a field experiment. We approach it with the same methodology we'll use for any other sports data science project. Being organized about our tools is part of the success.

It's interesting to observe the evolution of tools for data science. My dad works at Fermilab as a physicist specializing in linear accelerators, and occasionally he writes for physics magazines. For many years he's been using the same old version of numerical tools because they were so fine-tuned to his experiments and writing process. So, he turned to virtual machines (VMs) to keep the same version of the operating systems and tools and quickly became very efficient in using and managing VMs.

Virtual machines are kind of heavy. We know of many better ways to keep things isolated these days: virtual environments or containers. So, before we start diving into tools for machine learning and experimentation, I'd like to spend a few moments discussing virtual environments.

Virtual Environments

I often work on many things at once: experimentation is a natural part of data science. It didn't take me long to realize that I needed a way to keep things organized, as the warning signs of tsunami of dependencies started to appear: different versions of Python and various versions of deep learning tools with intricate dependencies. With all the versions of libraries and toolkits, things quickly started to get confusing and my humble laptop needed an upgrade. Before the next data science project unleashed another named tropical storm upon my cozy world, I knew I had to start working smart.

Think of a virtual environment as a sandbox (Figure 3-3) for your data science projects. They are a great way to work and keep things organized, because of the isolation needed in multiple projects, with multiple tools and dependencies. Traditionally, in Python ecosystem, there are several ways to create a virtual environment; the most commonly used are virtualenv, pipenv, and conda. Conda, the command-line abbreviation for a data science–centric Python distribution called "Anaconda," appeals to many data scientists because it's so easy to start with, and it has a great deal of data science tools built-in. In this book I use Anaconda, but the idea is similar to the others and I highly recommend trying any of these virtualization tools!

Figure 3-3. Virtual environments provide a sandbox when working with multiple projects and libraries

As a rule of thumb for each project, it's a good idea to create a separate virtual environment. To create a blank virtual environment in Anaconda, simply try this, substituting env-to-create with the name of your environment:

```
$ conda env create --name env-to-create
```

To list environments, use:

```
$ conda env list
```

For projects in this book, I included some environments that you can clone, with a typical set of requirements and dependencies. You'll find Anaconda environment templates in the source code as .yml files. When using pip, the standard way to keep track of dependencies is by using requirements.txt. To create a project environment when dependencies file is provided, try this:

```
$ conda env create -f environment.yml
```

The following are some useful Anaconda commands you can use to manage virtual environments:

```
# Remove an environment
$ conda env remove --name env-to-remove

# Create an environment with a specific version of Python
$ conda create --name env-to-create python=3.7

# Create an environment with specific modules
$ conda create --name env-to-create numpy matplotlib

# Activate an environment
$ conda activate env-name

# Deactivate an environment, back to prompt
$ conda deactivate

# Install a package from within an active environment
(conda-env) conda install tensorflow-gpu

# Export environment file
(conda-env) conda env export --file environment.yml
```

Figure 3-4. Notebooks provide a natural experimentation environment for a data scientist

Notebooks

I realized this book needs a special section on notebooks, because, frankly, they are so cool and useful for any data scientist (Figure 3-4)! I can't think of anything that makes a data scientist's day more pleasant, apart from a morning cup of coffee, than opening a fresh instance of Jupyter Notebook and adding the first cell of markdown, reflecting on the research to come. Your Jupyter notebooks may include markdown text, code, as well as images and rendering of visualizations, charts, and even video.

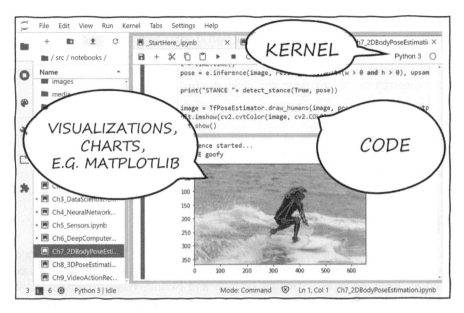

Figure 3-5. Jupyter notebooks integrate coding in many languages, rich text with markdown, mathematical notation like LaTeX, visualizations, and more

Jupyter notebooks (Figure 3-5) started with IPython project in 2014, the name coming, perhaps, from Galileo notebooks discovering moons of Jupyter. Besides Python, notebooks also support R and Julia, languages that became very popular in data science: even the name Jupyter may actually an abbreviation from *JU*lia, *PYT*hon, and *R*, but other languages are also supported, although a little more work and patience may be needed to configure things right.

Notebooks to data scientists are what sketch books to artists: they are ideal for data science projects in so many ways. They let the thought of a data scientist flow easy, telling the story of the project rather than being a rigid structure of files in traditional programming development environment (IDE). Some data science notebooks are as fun to read as a novel. They express the thinking process; you can read them as a journey, a book, a story of research. Yet, unlike fiction, they don't let the author deviate much from the science, supported by embedded code, data, and charts. Like any good science, a good notebook flows like a story, following the scientific approach: collecting empirical evidence, and forming and testing the hypothesis.

Perhaps, scientific books and articles of the future become more like Jupyter notebooks: instead of static text, they become active, with code and charts, and open to evolve research and experimentation. Most projects in the book are also in the form of notebooks, so read on for some useful tips on using them in applied data science!

As you get deeper into the data science, you'll realize that notebooks can also be difficult to operationalize in the "real-world" data science projects, especially when you need to scale your development, work in a team, or deploy a real operationalized service. Check the last chapters of this book to see how to take your research to the next step and make it a real data science product.

■ **Note** Notebooks are a great tool for research, but to take your data science project to production level, check the last two chapters of this book discussing strategies for automation and continuous deployment.

Markdown, Text, and Math

When you type in Jupyter notebooks, you can create "cells" that contain either code or markdown text. Markdown is a lightweight formatting syntax used in many websites, including GitHub and Reddit. For mathematics and scientific notation, LaTeX historically has been the standard, and notebooks provide full support for it. You can mix and match, depending on what you're trying to convey. For example, Newton's second law of motion can be written in markdown cell as follows, with %%latex magic (we will discuss "magic" command in the following sections):

```
%%latex
\begin{align}
F=ma
\end{align}
```

or in a code cell:

```
from IPython.display import display, Math, Latex display(Math(r'F = ma'))
```

Notebooks in the Cloud

While running data science projects on your local computer may be sufficient for some projects, running notebooks in the cloud may be necessary for many reasons (Figure 3-6). Some notable examples involve collaborating on developing models as a team, scaling models and compute capacity, and making your models available to end users with apps and APIs. For a deep dive into cloud-based machine learning, check Chapters 11 and 12 at the end of this book with practical examples on developing, training, storing, and consuming models in the cloud. Fortunately, getting started with machine learning in the cloud is easy, and you have many options: Microsoft Azure, Google Colaboratory, Amazon, and others. There's typically a free account that allows you to do some experimentation, and more with extended paid options.

Figure 3-6. Notebooks in the cloud

Notebooks require a kernel to execute code blocks. Kernels in Jupyter are execution environments, supporting many languages in addition to Python, for example, there are R, Julia, JavaScript, and more kernels available. To execute some code in a notebook in the cloud, it needs to be connected to a cloud compute instance. Often, we only need to glance at the notebooks, using cached execution results, for example, when looking at notebooks stored in GitHub. Looking at raw notebook files without some support of visualization would be rather difficult, so source repositories like GitHub or Azure DevOps (with extensions) provide basic code rendering of Jupyter files. In other words, even without code execution with these notebook servers, you can still see static notebooks with embedded data, graphics, and code highlighting, using cached results.

■ **Note** A great feature of a notebook is its ability to keep state and data, and you should be careful about it for the same exact reason! Too much data embedded in the notebook makes it huge and hard to manage, so you should know what you are doing.

Figure 3-7. Magic notebook commands

Notebooks Magic

Notebooks include many "magic" commands that you can use to accomplish various tasks that don't necessarily belong in the kernel (Figure 3-7).

With each release, the number of commands grows, so you can access the latest list with %lsmagic, which gives you a complete list of magical things:

```
# List magic commands in Jupyter
%lsmagic
```

For example, you can use %run magic command to execute external Python scripts stored outside of the notebooks as files. Another useful bit of magic for timing is %timeit. Timing scripts may become important to estimate various aspects of model training, loading data, and performance measurement, and notebooks provide convenient magic for that:

```
%%timeit
"-".join([str(n) for n in range(100)])
19.8 µs ± 2.53 µs per loop (mean ± std. dev. of 7 runs, 10000 loops each)
```

This script shows how %timeit magic command is used to measure the timing of a statement. Some other useful commands such as %writefile are used to write the contents of a cell to a file, which may be useful if you are generating scripts based on your notebooks, for example, scripts for model training. You'll see examples of this magic in the following chapters explaining machine learning automation.

Setting Up and Starting Notebooks

Installing Jupyter is super easy once you have Anaconda:

```
# Install and launch Jupyter notebook
$ conda install jupyter
$ jupyter notebook
```

In addition to classic Jupyter notebooks, I also recommend JupyterLab, a more modern environment that helps keeping track of multiple assets that inevitably become part of your research: files, images, video, and so on. JupyterLab is built on top of Jupyter and provides an enhanced interface to the familiar notebook:

```
# Install and launch JupyterLab
$ conda install jupyterlab
$ jupyter lab
```

Once launched, a URL pointing at the Jupyter server will open in your browser, and you are good to go. Once in Jupyter, you can use any of the *kernels* installed on your machine. One of the things that makes Jupyter so powerful is the ability to support multiple kernels. You can pause, resume, stop, and restart kernels as needed.

Once you create a conda virtual environment, and start working with notebooks, you may also want to add it as a kernel to Jupyter:

```
# Add conda environment as a kernel to Jupyter
$ python -m ipykernel install --user --name conda-env --display-name "Python
(conda-env)"
```

Data

Even the name data science presumes that we work with data a lot! Two data-related libraries that are essential in most data science projects in Python are NumPy and pandas, and there're many others that you'll need for specific data-related tasks. In machine learning data transformations are important to get a dataset ready for model training, for example, by manipulating and normalizing datasets.

NumPy

NumPy is one of the most fundamental scientific libraries in Python, representing multidimensional arrays. In data science NumPy is ubiquitous and provides internal data representation for most datasets: in addition to fast performance, it provides efficient operations with indexing, slicing, statistics, and persistence.

To illustrate, one of the features of NumPy we'll use a lot throughout this book is its ability to load and manipulate images as arrays. Most image classification methods you'll encounter take advantage of this NumPy capability. Each image is represented as a three-dimensional array of height, width, and color. The following code snippet loads an image of a jumping kiteboarder, then converts it to NumPy array using np.array() method, and uses shape property to display the size of the image. Note that the NumPy image also includes image channels as the third dimension:

```
from PIL import Image
import numpy as np

# Load image using PIL
im = Image.open('media/kiteboard_jump.jpg')
# Make a numpy array
arr = np.array(im)
# Print shape (size)
print('shape:',arr.shape)

Output:
shape: (512, 512, 3)
```

We can use NumPy slicing and indexing to quickly select areas of the image, by simply specifying rows and columns in the array; like in the following sample, the resulting cropped image is 300x300 pixels:

```
kiteboarder = arr[:300,100:400]
print('new shape:',kiteboarder.shape)

Output:
new shape: (300, 300, 3)
```

Next, let's display both images using Matplotlib methods (Figure 3-8):

```
import matplotlib.pyplot as plt
%matplotlib inline

fig, ax = plt.subplots(ncols=2)
ax[0].imshow(im)
ax[1].imshow(kiteboarder)
plt.show()
```

Figure 3-8. Loading and manipulating images with NumPy

Data Modeling and Pandas

NumPy provides powerful multidimensional array operations, and you can also persist and load data from arrays, using save() and load() methods, but to work with other data formats, such as Excel or CVS, and model data in tabular form, we can use another library called pandas, which stands for Python data analysis library (Figure 3-9).

Figure 3-9. Pandas stands for Python data analysis library

Pandas library is built on top of NumPy; it provides several data structures that turn out to be very useful in data science: Series and DataFrames. DataFrames object is what makes pandas very useful in our research; it adds standard operations on rows and columns of tabular data, grouping and pivoting. In the following example, here are some stats from my Active Fitness app (http://activefitness.co), showing grouping of trails that people track with the app, by activity season summer or winter:

```
d = {'category':['summer','summer','summer','summer','winter', 'all-season',
'summer', 'winter','winter','summer', 'summer'],
'sports':['running','walking','cycling','hiking','alpine ski', 'dog
walking', 'mountain bike', 'cross country ski', 'snowboard', 'inline
skating', 'trekking'],
'trails':[323909,174189,85908,56751,40339,33573,9957,8817,8674,5495,3051]
    }
df = pd.DataFrame(d)
print(df)
```

	category	sports	trails
0	summer	running	323909
1	summer	walking	174189
2	summer	cycling	85908
3	summer	hiking	56751
4	winter	alpine ski	40339
5	all-season	dog walking	33573
6	summer	mountain bike	9957
7	winter	cross country ski	8817
8	winter	snowboard	8674
9	summer	inline skating	5495
10	summer	trekking	3051

As you can see, from about a million trails tracked by users of Active Fitness platform, most are summer activities: running, walking, and cycling; followed by alpine ski as the most popular winter activity. Using pandas, we can quickly show statistics for a dataset, by using describe() method:

```
df.groupby('category').describe()
```

Trails	count	mean	std	min	25%	50%	75%	max
category								
all-season	1.0	33573.000000	NaN	33573.0	33573.0	33573.0	33573.0	33573.0
summer	7.0	94180.000000	118257.661585	3051.0	7726.0	56751.0	130048.5	323909.0
winter	3.0	19276.666667	18240.655864	8674.0	8745.5	8817.0	24578.0	40339.0

In the following section, we'll discuss visualization tools and libraries such as Matplotlib. A natural next step in visualizing our results is creating a plot of our data. For example, to plot the DataFrame of activities by number of users in Active Fitness, we can do it in just a few lines of code with Matplotlib (Figure 3-10):

```
import matplotlib.pyplot as plt
df.plot(kind='bar', x="sports", y="trails")
plt.show()
```

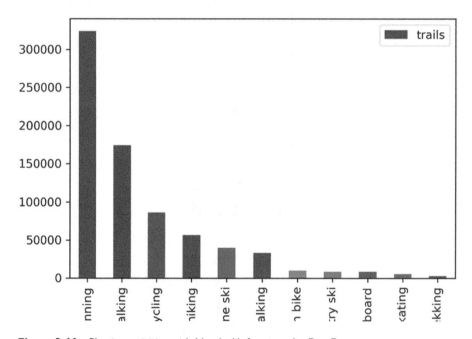

Figure 3-10. Plotting activities with Matplotlib from pandas DataFrame

Visualization

In data science we use many tools and libraries to analyze and visualize datasets. It is often important to convey results of our research to those who may not necessarily have the deep knowledge of our data, and perhaps, many times you've heard that many people are "visual learners." In sports coaching, we often use VAK (visual, auditory, kinesthetic) model to describe different learning styles. Not surprisingly, most humans are visual learners, because our brain's neural nets work very efficiently analyzing visual information. Fortunately, we have lots of tools, so let's get started!

Figure 3-11. Tools like Matplotlib provide a way to visualize your experiments

Matplotlib

Originally created by John D. Hunter, a neurobiologist, Matplotlib library is one of the most widely used tools in modern data science (Figure 3-11). To illustrate some of the concepts and use of visualization tools, let's look into medical imaging data, such as magnetic resonance imaging (MRI). I used an example of data scanned from my right knee, taken after I had a ligament sprain lightly when I was skiing in Sochi. For a practical example, let's load one of the scans of my knee, using pydicom, a library that can read MRI data and play with the results in Matplotlib. If you haven't installed pydicom, you can do it easily with pip right from your notebook or command shell:

```
$ pip install pydicom
```

You can then import the library into your notebook as follows, and then load one of the files with dcmread method:

```
import pydicom as dicom

ds = dicom.dcmread(dicom_file)
print(ds.StudyDescription)

Output:
Knee R
```

This library also includes anonymized MRI datasets, so we'll have plenty of data to play with for our visualization endeavors. Remember, data science is all about data! To import the NumPy module and display the size of the image array we just loaded earlier from the MRI scan, simply run the following code. Notice that the pixel_array property from the preceding example also gives us a NumPy array representing the image, no further conversion necessary! Conveniently, Matplotlib includes even a color map, which is a representation of a color space, for bone tissue (plt.cm.bone) to display our scans; now it looks like an actual scan from a medical office (Figure 3-12):

```
import matplotlib.pyplot as plt
%matplotlib inline

plt.imshow(img, cmap=plt.cm.bone)
plt.show()
```

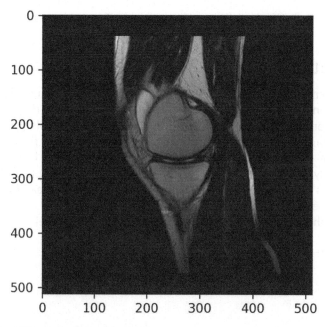

Figure 3-12. MRI scan loaded in Matplotlib

As data scientists, we'd like to study our scan to see if we can perhaps isolate regions on it, for example, to identify features or areas of interest, so we create what's called a binary mask (Figure 3-13), an image that contains only black and white channels. This type of image transformation is often used in computer vision and tasks like classification and semantic segmentation. First, we fill an array of the same shape as our image with zeros and then highlight everything we need relative to the mean of the array:

```
arr = np.zeros(img.shape)
mask = img <= img.mean()
arr[mask] = 1
mask = img > img.mean()
arr[mask] = 0
plt.imshow(arr, cmap=plt.cm.bone)
plt.show()
```

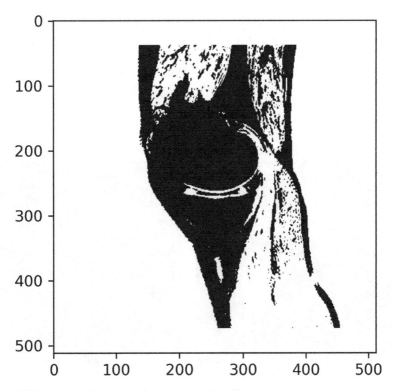

Figure 3-13. Applying binary mask, using basic NumPy

SciPy, scikit-image

While NumPy provides basic array manipulations and, as you can see, allows implementing some image transformations, you may need other tools for more advanced methods and algorithms. SciPy ecosystem of scientific libraries for Python provides a wealth of those. Let's say we want to detect an edge on the MRI scan image we just loaded: detecting contours is often needed for machine learning to label data and have models trained on specific features of the image. In computer vision, Sobel, Roberts, or Canny operators are often used for edge detection, and fortunately scikit-image contains most of these methods built-in (Figure 3-14). In addition to edge detection, scikit-image also provides methods for image segmentation (Figure 3-15) along with many other deep vision related helpers:

```
from skimage import filters
edge_sobel = filters.sobel(arr)
edge_roberts = filters.roberts(arr)
fig, ax = plt.subplots(ncols=2, sharex=True, sharey=True)
ax[0].imshow(edge_sobel,cmap=plt.cm.bone)
ax[0].set_title('Sobel Edge Detection')
ax[1].imshow(edge_roberts, cmap=plt.cm.bone)
ax[1].set_title('Roberts Edge Detection')
plt.tight_layout()
plt.show()
```

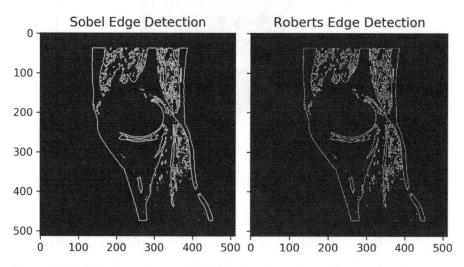

Figure 3-14. Edge detection using scikit-image

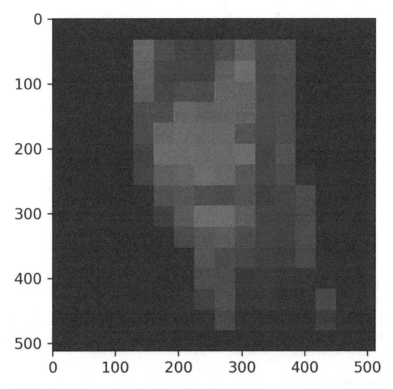

Figure 3-15. Using image segmentation with scikit-image

PROJECT 3-1: USING ACTIVE CONTOURS SEGMENTATION

Sport movement analysis often requires isolation of body parts, and segmentation methods like active contours provide the first step in that direction. Let's see how we can use active contours to select a part of athlete's body. In the following snippet, we'll use Python Imaging Library (PIL) to load this image and convert it into NumPy array (Figure 3-16):

```
from PIL import Image
image = Image.open('data/images/skiers.jpg')
img = np.array(image)
plt.imshow(img)
plt.show()
```

Figure 3-16. Original image loaded into NumPy array and visualized with Matplotlib

Next, we use scikit-image to initialize a circular area and let snakes `active_contour` method fit the spline to lines and edges on the image of the athlete (Figure 3-17). This approach is supervised, because we need to define the initial region.

```
import numpy as np
import matplotlib.pyplot as plt
from skimage.color import rgb2gray
from skimage import data
from skimage.filters import gaussian
from skimage.segmentation import active_contour

# Make image gray tone for contours
img = rgb2gray(img)
# Initialize spline
s = np.linspace(0, 2*np.pi, 200)
init = 80*np.array([np.cos(s), np.sin(s)]).T + 250
# Fit spline to image
snake = active_contour(gaussian(img, 3),init)

fig, ax = plt.subplots(dpi=150)
ax.imshow(img, cmap=plt.cm.gray)
```

```
ax.plot(init[:, 0], init[:, 1], '--r', lw=3)
ax.plot(snake[:, 0], snake[:, 1], '-b', lw=3)
ax.axis([0, img.shape[1], img.shape[0], 0])
plt.show()
```

Figure 3-17. Result of segmentation with active contours method

In this example, we used a circle to provide an area for active contours algorithm. In more sophisticated scenarios, methods such as GrabCut, developed and open sourced by Microsoft researcher, could be used to isolate segments, based on multiple masks.

OpenCV

OpenCV has been an important part of computer vision toolset and includes, besides core image and video processing, 2D and 3D methods for features and reconstruction, deep neural networks (DNN), and learning modules.

A couple of years ago, during a hackfest organized by Microsoft Garage, we were presented with a very interesting problem: detecting motion of an Olympic diver, combining data from the springboard and after the takeoff in the air. Working with the team of interns and guided by Phil Cheetham, from the US Olympic Committee, in three days we were able to implement a solution, dividing the task between a flex meter IMU sensor installed on the springboard, and using OpenCV to track the diver movement in the air. We published this research in Microsoft Developer Magazine, entitled "Machine Learning – Analyzing Olympic Sports Combining Sensors and Vision AI."

The method we used for the air part of the movement is called optical flow; it's used in computer vision, robotics, and is built into some common sensors, such as computer mouse. The idea behind optical flow is that intensity (or brightness) of an object between two frames, assuming movement is small, doesn't change, and this assumption is used to detect the point in the next frame, making it possible to track points of the video over time.

For a data scientist working with video and image data, OpenCV provides many useful methods: object detection, tracking, segmentation, and more. In the source code notebook for this chapter, I provided an example of using optical flow and OpenCV.

Deep Learning Frameworks

In this book we use some of the most popular frameworks available in data science today: PyTorch, Keras, and TensorFlow for deep learning projects. For reinforcement learning, you will also encounter examples with OpenAI Gym and some other libraries, including physics for simulation. Some of these frameworks started as internal projects at Google, Facebook, Uber, and Microsoft and were made open source eventually. Most of data science projects benefit from accelerated hardware, so when you install these libraries, make sure you are using a graphics processing unit (GPU)-accelerated version, as they offer a significant performance boost!

PyTorch

> *I've been using PyTorch a few months now and I've never felt better. I have more energy. My skin is clearer. My eyesight has improved.*
>
> —Andrej Karpathy

PyTorch is an open source machine learning framework originally based on Torch, and largely developed by Facebook AI research lab. In recent years PyTorch gained popularity among data scientists and features NumPy-based tensor operations with `torch.Tensor` object. At the low level, PyTorch uses an automatic differentiation, called autograd system to take gradients, nn module for higher-level neural networks. As you may recall from the chapter on neural networks, learning is based on backpropagation, and PyTorch provides a very efficient implementation that automates this process. As a fully featured deep learning library, PyTorch also has modules for data augmentation and loading.

Many examples in this book are in PyTorch, including video action recognition, as well as classification examples. One interesting feature of PyTorch is the module called torchvision, which includes many models for deep vision, including automatically fetched pretrained models. Since this book is about applied practical methods, taking advantage of pretrained models for transfer learning helps a lot!

TensorFlow

TensorFlow is another popular open source framework, originally developed by Google Brain team. TensorFlow is very popular for many reasons: it has a huge support from the open source community, as well as Google AI team. Building models at high level with TensorFlow is easy with another open source library called Keras. TensorFlow has interesting features, including ability to run in JavaScript environments and mobile and edge devices. For example, body pose estimation example explained further in this book can run near real-time in your browser with TensorFlow.

Another great feature of TensorFlow is TensorBoard, a visualization toolkit that includes stunning visualization methods, including 3D visualizations a practical data scientist specializing in sports can use for human pose analytics with plug-ins like mesh.

Keras

If you don't use Keras, and you are starting your journey as a data scientist, you probably should give it a try! Keras is especially popular at making model prototyping easy. Here's how easy it is to make a multilayer model with Keras:

```
from keras.models import Sequential
from keras.layers import Dense

model = Sequential()

model.add(Dense(units=64, activation="relu", input_dim=10))
model.add(Dense(units=10, activation="softmax"))
```

Reinforcement Learning

Reinforcement learning (RL) is learning what to do by trying, and taking the next action based on the reward. In Chapter 10, "Reinforcement Learning in Sports," of this book, which provides a deep dive into this method, you'll see many practical examples of using reinforcement learning for sports, health, and fitness applications.

OpenAI Gym

Reinforcement learning toolkits like OpenAI Gym are popular with data scientists, and in this book, you will find several OpenAI Gym examples. In addition to the environment, many libraries have emerged recently that include methods and algorithms for reinforcement learning: Deep Q Networks (DQN), Deep Deterministic Policy Gradients (DDPG), Generative Adversarial Limitation Learning (GALL), and various implementations of actor-critic (A2C). OpenAI has been created as a simulation environment, a playground to unfold the "scene" of the action and train models. It remains open for research and at what algorithms should be employed to solve actual problems, with main areas being games, 2D physics problems on Box2D, classic control physics, 3D physics, and robotics, such as PyBullet and text.

Cloud, Automation, and Operationalization

You'll see the term CI/CD (continuous integration/continuous delivery) many times referring to the development cycle in machine learning, and in the last chapters of this book, we'll be using some practical examples and cloud-based tools for taking your research to the level of best practices and standards used in modern data science.

Summary

This chapter provides an overview of some of the basic tools you may need as a data scientist in health and fitness. In the next chapters, we'll dive into some practical examples of building, training, and consuming models for various tasks in health and fitness, using these and many other deep learning tools and technologies.

As I was working on this chapter, I had an inspiring call with Dr. Phil Cheetham, an Olympian in gymnastics, a sport scientist on US Olympic Committee, and an innovator in biomechanics. Biomechanics is a constantly evolving field, and machine learning adds a completely new aspect to it. Phil inspired me to do this work to encourage the next generation of sport scientists to learn new AI-based methods, using new tools and techniques. As I write this, I keep thinking about Phil and how I can make this useful, so he and other data scientists can apply these tools in their practical research. In the chapters that follow, as you read about practical data science applications in health and fitness, check out the source notebooks with examples and projects. Most of them will be using tools I mentioned in this chapter.

Neural Networks

There's something magical about Recurrent Neural Networks (RNNs).

—Andrej Karpathy

Defining a Neural Network

Neural nets are magical: I remember training my toy JetBot Nano robot with 100 "blocked" and "unblocked" images, and the data science toy started intelligently avoiding spots in my house. The neural net inside the bot learned to do a collision detection based on relatively small dataset I gave it to train, and it truly worked like magic! If you are an engineer used to conventional rule-based programming, just thinking of collision situations, with obstacles being of different shapes, taken at different lighting, different angles, will make your head spin. How long would it take to code such a complex system without neural nets? When you learn a new language, you learn grammar rules. I barely remember the names of grammatical terms, yet somehow the neural net inside my brain pulls the right combination of words in the order defined by very complex grammatical rules I have no idea about. As you get better and your brain infers the grammar, it becomes easier for you to apply these rules. It seems magical how the neural network inside the brain works: the magic of learning just happens. But how exactly does it happen?

© Kevin Ashley 2020

K. Ashley, *Applied Machine Learning for Health and Fitness*,
https://doi.org/10.1007/978-1-4842-5772-2_4

Neurons

Neural nets are best explained with a biological analogy, the neurons. Basically, a neuron has multiple input connections, called *dendrites*, that receive synaptic inputs, and one output called *axon*. The strength of synapses indicates weights *w* of input signals *x*. In the neural cell, called *soma*, input signals are summed up, and if the strength exceeds a certain threshold, it fires an output signal via the axon output. The idea of creating a mathematical model to describe neural nets comes back to 1943, with the McCulloch-Pitts model. In the following years, it was refined: Frank Rosenblatt (1958), and Minsky-Papert further extended this model in the 1960s, with the introduction of **weights** as the measure of importance and the mechanism of **learning** those weights (Figure 4-1).

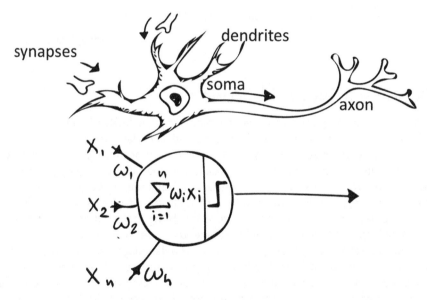

Figure 4-1. Neuron and its mathematical model

Our mathematical neuron has multiple inputs *x* with weights *W*. The work done by soma mathematically translates to a summation or a *dot product* of inputs, plus a value of the bias:

$$\sum_{i=1}^{n} W_i x_i + b_i = W \cdot x + b$$

After the summation, an activation function is used to fire the output, if the sum is over a certain threshold. So in the most basic form, the model of a neuron can be written as:

$$f(x) = f(W \cdot x + b)$$

Activation is the term that defines the mathematical gate that turns neuron "on" or "off". In our neuron model, $f(x)$ is the activation function, or nonlinearity, given the dot product of weights $W \cdot x$ and a bias b. Read on the following section, for activation functions in more detail.

Activation

Activation can technically be any function triggering output after a certain threshold input value (Figure 4-2). In fact, there're several choices for activation functions; the most popular options are sigmoid, hyperbolic tangent, and ramp function:

- Sigmoid: $\sigma = 1/(1 + e^{-x})$

- Hyperbolic tangent: $tanh = (e^{2x} - 1)/(e^{2x} - 1)$

- Ramp function or rectified linear unit (ReLU): $f(x) = max(0, x)$

Figure 4-2. Activation function "activates" the output once the sum is above the threshold

To visualize the difference, let's plot these functions (Figure 4-3):

```
import matplotlib.pylab as plt
import numpy as np
%matplotlib inline

x = np.arange(-4,4,0.01)
plt.plot(x, 1/(1+np.exp(-x)),'k--', label="sigmoid")
plt.plot(x, np.tanh(x),'k:', label="tanh")
plt.plot(x, np.maximum(x, 0),'k', label="relu")
plt.xlabel('x')
plt.ylabel('f(x)')
plt.legend()
plt.show()
```

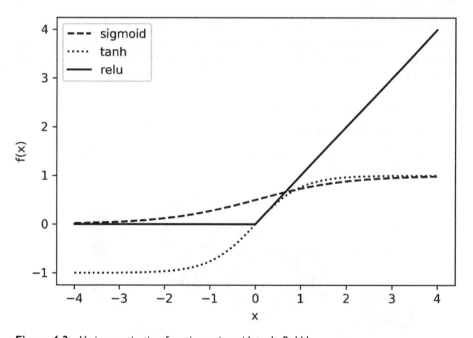

Figure 4-3. Various activation functions: sigmoid, tanh, ReLU

It's easy to see why rectifier and hyperbolic tangent functions are popular for activation: ReLU has a threshold at 0, and tanh is zero centered, returning values in the range [-1,1]. Sigmoid is a classic example of the activation function, it returns values in the range [0,1], and it is not as steep as tanh and ReLU.

Perceptron

Perceptron is a binary classifier used in supervised learning. The original Perceptron was a pretty heavy machine by Frank Rosenblatt, built at Cornell University and funded by the US Office of Naval Research. The name comes perhaps from the idea that at some point the machine can learn to recognize images. After half a century of AI research, we now have multilayer neural networks that can do it easily, but you'll be surprised how many ideas are coming from the original perceptron!

■ **Note** While there're many machine learning frameworks out there, I recommend stepping through notebooks in this chapter that create neural networks from scratch, with pure Python and math. This helps understanding the principles of neural nets, while frameworks may abstract the details.

In this project, we will use a model similar to a single-layer Perceptron, to create a single-layer neural network. Later in this chapter, we'll look into deficiencies of such a model, and what can be done to improve it.

> ## PROJECT 4-1: BUILDING AND TRAINING PERCEPTRON
> ## FROM SCRATCH IN PYTHON

As part of this project, let's build the model in pure Python, from scratch, without using any fancy libraries or tools to illustrate the concepts of neural nets. Perceptron is an example of what's called **supervised learning**, because we provide a set of labels, or expected outputs in order to train it:

```python
class Perceptron():

    def __init__(self, features):
        self.weights = np.zeros((features, 1))
        self.bias = 0

    def activation(self, x):
        return np.where(x>=0, 1, 0)

    def predict(self, x):
        return self.activation(np.dot(x, self.weights) + self.bias)

    def train(self, inputs, labels, eta=0.1, epochs=10):
        errors = []
```

```
for t in range(epochs):
    # Calculate prediction
    prediction = self.activation(np.dot(inputs, self.weights) +
    self.bias)
    # Adjust weights and bias
    self.weights += eta * np.dot(inputs.T, (labels - p
    rediction))
    self.bias += eta * np.sum(labels - prediction)
    # Calculate mean square error
    mse = np.square(np.subtract(labels,prediction)).mean()
    errors.append(mse)
    print(f"epoch {t}/{epochs} mse: {mse}")
```

In just a few lines of code, our Python perceptron object implements the artificial neuron, following the models created in the middle of the last century.

Creating a Dataset

Can a neural network containing a single neuron solve logical problems? For our project let's use a logical operation, such as binary AND gate, which is the basis of most logical/rule-based programming and digital circuits.

■ **Note** Other logical operations include OR, NOT, and so on. One operation we do not want to test *yet* is XOR for the reasons I'll explain in the following pages. Let's just say for now that a single perceptron is unable to solve XOR because it's not linearly separable.

Our logical AND operation should resolve to the following values:

```
1 AND 1 = 1
1 AND 0 = 0
0 AND 1 = 0
0 AND 0 = 0
```

From these rules, I'll make an array from the left-hand side of the equation and provide it as an *input* to perceptron:

$$(x_1, x_2)=(1,1),(1,0),(0,1),(0,0)$$

Our *labels* (values we expect perceptron to predict) come from the right-hand side of the equation: (1, 0, 0, 0). Let's initialize our dataset, with two NumPy arrays:

```
# Initialize training dataset
labels = np.array([ [1], [0], [0], [0]])
inputs = np.array([[1, 1],[1,0],[0,1],[0,0]])
```

Initializing the Model

The perceptron class is initialized with weights and bias. The number of features (0 and 1) is the dimensionality of the dataset we intend to infer:

```
def __init__(self, features):
        self.weights = np.zeros((features, 1))
        self.bias = 0
```

Initial weights and bias   By default, our neuron is initialized with zero weights and bias. It is often recommended to initialize weights with random small numbers.

Training the Model

The train method is what we use to make the model learn the weights it needs to predict real inputs. The goal of training for the model is minimizing loss, or difference between actual observation and prediction, therefore improving the model's accuracy (Figure 4-4). In the training function, we supply several *hyperparameters*; in machine learning speak, this is a set of parameters that are set before learning begins. Training happens over a series of iterations we call *epochs*. Another hyperparameter we pass to train method is *learning rate*. Learning rate is the step size of training; it controls the speed of convergence. Our activation function for this binary neuron is defined as np.where(x>=0, 1, 0) and returns 0 or 1:

```
def activation(self, x):
        return np.where(x>=0, 1, 0)
```

Figure 4-4. Training the model

Just as we expressed earlier in mathematical terms, the summation is calculated as a dot product of existing weights, plus the bias. We call the activation function and pass the summation to it:

```
prediction = self.activation(np.dot(inputs, self.weights) + self.bias)
```

Now, I can create an instance of my perceptron object and train it, by calling train() method and passing inputs and labels to it:

```
perceptron = Perceptron(2)
perceptron.train(inputs,labels)
```

In the Perceptron object, I used 10 epochs as a default, and the training converges very quickly (Figure 4-5):

```
epoch 0/10 loss: 0.75
epoch 1/10 loss: 0.25
epoch 2/10 loss: 0.25
epoch 3/10 loss: 0.0
epoch 4/10 loss: 0.0
epoch 5/10 loss: 0.0
...
```

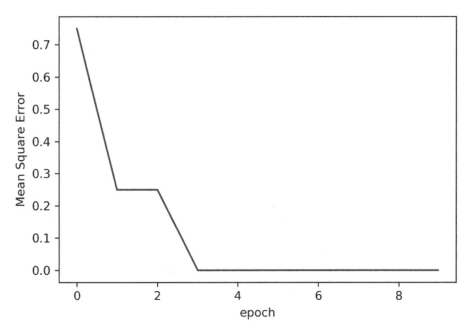

Figure 4-5. Convergence of the perceptron model

Note that in the preceding example, we also calculate the loss, also known as a cost function. For cost, we can use mean squared error (MSE) or cross-entropy. The cost becomes small when the predicted values get closer to the expected output, so we can write this down as:

$$C = \frac{1}{n}\sum_{i}^{n}\left(x_i - y_i\right)^2$$

or in Python:

```
loss = np.mean((labels-z[1])**2)
```

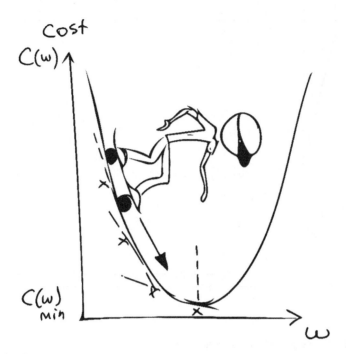

Figure 4-6. Gradient descent

The goal of neural network is minimizing the cost C. The method of minimizing the loss often implies using *gradient descent* to find the minimum of the cost function (Figure 4-6).

Validating the Model

The perceptron learns very quickly (our dataset is very small); let's give it a test value to see if it predicts the result correctly:

```
# Testing prediction
test = np.array([1,1])
print(perceptron.predict(test))
```

```
Output:
 [1]
```

For a logical gate, we chose a very small dataset, containing only a few values. Let's test the Perceptron on a larger set of data. We'll make one exception here to our pure Python rule in this chapter and use scikit-learn and functions it provides for datasets. First, I use make_blobs function that conveniently generates random datasets we can further use for clustering (or Perceptron decision line separation). In this example, I defined two centers for the perceptron. I also used scikit-learn `train_test_split` function to split the dataset into test and training data:

```
from sklearn.datasets import make_blobs
from sklearn.model_selection import train_test_split
x, y= make_blobs(n_samples=200, centers=2, random_state=1)
x_train, x_test, y_train, y_test = train_test_split(x, y, random_state=1)
print(f'train: {y_train.shape} test: {y_test.shape}')
plt.scatter(x[:,0], x[:,1], c=y)
plt.show()
```

Now, I use the training dataset from the preceding code to train the perceptron on the training subset and display the weights and bias:

```
perceptron = Perceptron(2)
perceptron.train(x_train,y_train.reshape(y_train.shape[0],1))
print(f'w={perceptron.weights} b={perceptron.bias}')

Output:
weights=[[-19.44] [-40.50]] bias=-4.69
```

Decision Line

The perceptron we created works as a linear classifier (Figure 4-7) for data that can be separated linearly. The line separating two clusters of labeled data is called the decision line and is determined by the equation: $y = w_i x_i + b$. We *already learned* weights w_i and bias b; when we trained the model, they are now stored in `perceptron.weights` and `perceptron.bias` variables. Let's plot the decision line and verify that it separates both labeled clusters we created earlier (Figure 4-8):

```
x_ = np.linspace(x.min(),x.max(),100)
y_ = -(perceptron.weights[0]*x_ + perceptron.bias)/perceptron.weights[1]
plt.plot(x_, y_)
plt.scatter(x[:,0], x[:,1], c=y)
```

Figure 4-7. Linear classifier decision line

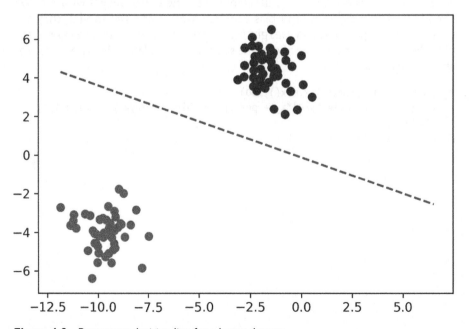

Figure 4-8. Perceptron decision line for a larger dataset

To summarize, our model correctly separates two larger datasets, and to achieve this, we used a simple linear classifier, confirming our assumption that these sets are linearly separable.

Multilayer Networks

It's easy for our Perceptron-dude to draw decision line for a dataset containing logical AND problem. But what if the data is not linearly separable (Figure 4-9)? The simplest case for such a dataset is logical XOR operation.

Figure 4-9. Multilayer perceptron is needed to predict composite problems, like XOR gate

As you can see, if we draw the XOR dataset next to AND dataset on the plane, the problem for a simple linear classifier, such as our perceptron, becomes apparent.

■ **What a single neuron cannot predict** A single-layer neuron works as a linear classifier, that is, it works when data is linearly separable (such as a logical AND, OR operations). One famous example where single-layer perceptron doesn't work that easily is a logical XOR operation.

Fortunately, problems like XOR gate can be represented as a composition of several logical operations:

$$x_1 \text{ } XOR \text{ } x_2 = \left(x_1 \text{ } OR \text{ } x_2 \right) AND \text{ } NOT \left(x_1 \text{ } AND \text{ } x_2 \right)$$

In terms of neural net, intuitively, it means we need additional *layers* to represent the composition. This is achieved, for example, by multilayer neural networks (Figure 4-10). The reason we call them networks is because these structures can be represented by graphs of composite functional transformations. A multilayer perceptron is also called a *feedforward* network because weights flow from inputs, through the neural network hidden layers, and to the outputs. Perceptron, when it has multiple layers, is a case of a feedforward network, and the idea that these networks can recognize images, dating back to the 1960s, is implemented in a special type of these networks called *convolutional neural networks*. Networks that in addition to feedforward connections include feedback connections are called *recurrent neural networks*.

Figure 4-10. Neural net graffiti: more complex classifiers with multilayer networks

Backpropagation

Can the brain do backpropagation?

—Geoffrey Hinton

I started this chapter with a biological analogy of neural networks. Backpropagation (Figure 4-11) is an algorithm that makes neural nets efficient at learning: most neural nets today employ backpropagation. But does this mechanism really exist in human brain? Geoffrey Hinton, the "godfather" of data science and author of backpropagation, says maybe, but many neural scientists disagree. There're many reasons why this mechanism, while very efficient in ANNs (artificial neural nets), may not work in nature in exactly the same way, while there could be other ways to optimize learning. Things that neuroscientists typically mention that make it hard for human brain to do backpropagation are the speed of synapses, the fact that synapses do not operate on real numbers (activation function either fires or not), and different values actually sent during forward and backpropagation in different directions. Geoffrey Hinton's lecture in Stanford, entitled exactly as the quote I brought forward in this section, points at various ways why backpropagation may still work in humans and also looks at other ways to optimize neural nets. In any case, for most ANNs, backpropagation seems like a good idea.

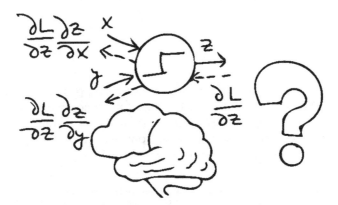

Figure 4-11. Can the brain do backpropagation?

In the previous examples of a feedforward network, we calculated loss of the network. Recall that the network can be represented by a graph, that is, a composition of functions: in *forward* propagation, the input layer takes the input weights, and each layer calculates the product and calls activation function *f*, which triggers the output. For a simple network, consisting of an input layer with weights W_0 and hidden layers with weights W_1, the *forward* pass is a function combining weights and biases of each layer.

We can write this down as a composition function, where f is the activation function, for example, a sigmoid, returning the *prediction* of each layer:

$$f(x) = f_1(f_0(x))$$

or in terms of weights and biases:

$$f(x) = f_1(W_1 \cdot f_0(W_0 \cdot x + b_0) + b_1)$$

In each layer of our network, we can calculate the error as cost function, that is, the difference between target values (or labels in data science speak) $\hat{y}$ and the layer prediction $y = f(x)$:

$$C = \frac{1}{2}(\hat{y} - y)^2$$

The derivative of this error by the predicted value is simply $\dfrac{dC}{dy} = \hat{y} - y$. For backpropagation, we calculate the derivative of the cost with respect to weights and biases of the network, calculated during the forward pass:

$$\frac{\partial C}{\partial W_i}, \frac{\partial C}{\partial b_i}$$

Backpropagation is fundamental to neural networks, providing a mechanism for learning. As I mentioned in the previous chapter, some tools like PyTorch include very efficient backpropagation methods, such as autograd system, built-in. We can argue if biological neurons use backpropagation, but it certainly helps mathematical neural networks train.

PROJECT 4-2: NEURAL NET WITH BACKPROPAGATION

In this project we'll take a logical XOR operation that our simple single-layer Perceptron wasn't able to solve, and make a neural net that can solve it with a neural network with multiple layers and backpropagation:

0 XOR 0 = 0

1 XOR 0 = 1

0 XOR 1 = 1

1 XOR 1 = 0

The goal of this project is to demonstrate that multilayer networks can solve nonlinear problems. Our network will include a hidden layer, in addition to input and output layers: in the constructor, we'll define network topology as layers=[2,2,1], where input and hidden layer are both shaped as inputs, for example, [[0,0],[0,1],[1,0],[1,1]], that is, both have two dimensions and the output layer is a one-dimensional vector, the same shape as the labels vector [0,1,1,0]. In the constructor we'll initialize weights and biases with random values in the range [0, 1]:

```
def __init__(self, layers=[2,2,1]):
    np.random.seed(1) # for consistency
    self.w = [np.random.uniform(size=(layers[i], layers[i+1])) for i in
    range(len(layers)-1)]
    self.b = [np.random.uniform(size=(1, layers[i+1])) for i in
     range(len(layers)-1)]
```

For backpropagation, we'll add both sigmoid and sigmoid derivative functions for activations:

```
def sigmoid (self, x):
  return 1/(1 + np.exp(-x))

def sigmoid_derivative(self, x):
  return x * (1 - x)
```

The forward function provides the forward pass of our network, calling activation on the dot product at each layer:

```
def forward(self, x):
 z0 = self.sigmoid(np.dot(x,self.w[0]) + self.b[0])
 z1 = self.sigmoid(np.dot(z0, self.w[1]) + self.b[1])
 return (z0, z1)
```

Backpropagation and weights update happen in backward method, for each layer we calculate gradients and pass them in the reverse order from the bottom layer of the neural net to the top:

```
def backward(self, z, labels):
    error = labels - z[1]
    d1 = error * self.sigmoid_derivative(z[1])
    d0 = d1.dot(self.w[1].T) * self.sigmoid_derivative(z[0])
    # Update weights/biases
    self.w[1] += z[0].T.dot(d1) * lr
```

```
        self.b[1] += np.sum(d1,axis=0,keepdims=True) * lr
        self.w[0] += inputs.T.dot(d0) * lr
        self.b[0] += np.sum(d0,axis=0,keepdims=True) * lr
```

In the training method, we iterate over a number of epochs through our neural network, first making a forward pass and calculating initial weights, then making a backward pass, calculating gradients, and updating weights based on the error. Finally, to calculate the overall cost and plotting it:

```
def train(self, inputs, labels, epochs=10000, lr=0.1):
    errors = []
    for t in range(epochs):
        z = self.forward(inputs)
        self.backward(z, labels)
        # Calculate loss (MSE)
        loss = ((labels-z[1])**2).mean()
        errors.append(loss)

    plt.plot(errors)
    plt.xlabel('epoch')
    plt.ylabel('loss (MSE)')
    plt.show()
```

To prove that our network works, let's supply it with the inputs and labels for our expected outputs and train the model:

```
# XOR operation:
# 0 XOR 0 = 0
# 1 XOR 0 = 1
# 0 XOR 1 = 1
# 1 XOR 1 = 0
inputs = np.array([[0,0],[0,1],[1,0],[1,1]])
labels = np.array([[0],[1],[1],[0]])

# Initialize and train our network
nn = NNBackprop(layers=[2,2,1])
nn.train(inputs,labels)
```

```
# Run prediction again to show the results
prediction = nn.forward(inputs)[1]
print(f'input: {labels.tolist()}')
print(f'prediction: {[[int(p > 0.5)] for p in prediction]}')
```

Results show that the model converges, and if we run the forward pass again, it returns expected results (Figure 4-12):

```
input: [[0], [1], [1], [0]]
prediction: [[0], [1], [1], [0]]
```

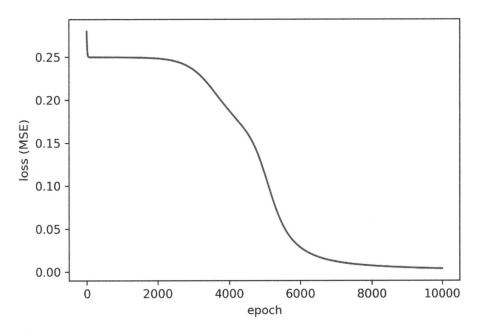

Figure 4-12. Convergence of the multilayer neural network with backpropagation

Summary

In this chapter I introduced the basics of neural networks and machine learning, starting with a biological analogy of a single neuron. As part of the practical projects, I created a fully functional model of Perceptron in Python, trained it with sample data, and visualized the results. Focusing on supervised learning, I demonstrated some of the shortcomings of a Perceptron model and described a more sophisticated neural net, containing multiple layers and backpropagation that can solve problems (such as XOR) that a single-layer network was unable to predict.

Sensors

Good science is good observation.

—Avatar

Figure 5-1. A kitesurfer riding the wave: neural net estimating body position despite highly dynamical environment. How's that possible? Read on

© Kevin Ashley 2020
K. Ashley, *Applied Machine Learning for Health and Fitness*,
https://doi.org/10.1007/978-1-4842-5772-2_5

Human motion is the subject of kinesiology and biomechanics, sciences that study mechanical aspects of human body movement (Figure 5-1), using many classical mechanics principles, such as kinematics and kinetics, which we covered in Chapter 2 when discussing physics of sports. In biomechanical analysis, *sensors* are devices that provide experimental data. We can use data collected from sensors to provide a high-quality ground truth data and train our models (Figure 5-2).

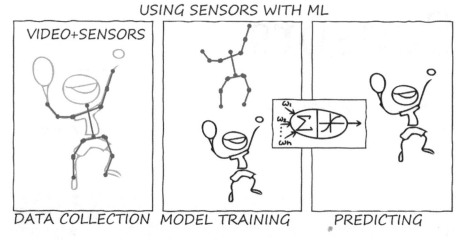

USING SENSORS WITH ML

VIDEO+SENSORS

DATA COLLECTION MODEL TRAINING PREDICTING

Figure 5-2. Connecting biomechanics, sensors, and machine learning

In this chapter, we'll do an overview of sensor technology and how to apply it for various biomechanical problems with pointers to sections in this book that provide practical projects on using data collected from these devices. Sensors also have very practical consumer applications: from helping athletes progress at sports to assisting coaches with movement analysis, providing feedback, logging, and recording data. The goal of any coach is to help athletes improve performance and prevent injuries: with machine learning methods, the possibility of a virtual sports and fitness assistant becomes reality.

■ **Features** A sport data scientist wish list often includes sensors that need to withstand harsh ocean water, snow, extreme temperatures range, be reliable and dependable, log data at highest possible frequency, connect to the cloud, provide mobility, ideally provide some on-device inferencing, and be inexpensive.

Although I use the term sensor, but the meaning is a broader set of IoT devices that include, besides the sensing unit (such as an optical sensor, a gyroscope, an accelerometer), a sensor processing unit (SPU), capable of computing and processing sensor data onboard. Similarly, other edge devices are, for example, TPU (tensor processing unit), VPU (vision processing unit), or HPU (holographic processing unit). The actual sensing devices are often called microelectromechanical systems (MEMS): most are inexpensive and can be easily integrated in custom boards.

Types of Sensors

Smart sensors that can collect data and act on physical events as they happen will enable us to build models that have knowledge about, and the ability to interact with, an incredible diversity of things in the physical world.

—Kevin Scott, Microsoft CTO, "Reprogramming the American Dream"

From many sensors a data scientist can use, Table 5-1 provides a map of the most commonly used devices.

Table 5-1. Mapping sensor devices to applications in health and fitness

Sensor	Function	Applications	Data output
Vision and cameras	Cameras are one of the most fundamental and useful type of devices for sport analysis. Since first capturing of still and moving pictures, used to record moving and still visual data. There're other optical devices, such as infrared cameras, that reflect light from markers.	Kinematics for most sport activities *Limitations: field of view, distance, complex ML*	60 fps x 4K x RGB channels
Inertial movement sensors (IMU, AHRS)	Integrate gyroscope, accelerometers, magnetometers, and optionally geolocation devices to provide information on kinematics of movement: acceleration, rotation, angular velocity, direction.	Kinematics with high frequency and precision: acceleration, rotation *Limitations: battery, body placement*	200Hz acceleration, rotation, angular velocity, magnetic force
Range imaging sensors	Various technologies to measure depth: time of flight, structured light, triangulation, interferometry, etc. (e.g., LIDAR illuminates subject with laser light and measures the reflected light that is used for 3D scanning of the scene).	Distance measurement *Limitations: range, field of view*	Distance, depth 10Hz

(continued)

Table 5-1. (continued)

Sensor	Function	Applications	Data output
Pressure	Pressure mats, force plates, and textile pressure sensors are used in sports and medical research to get pressure maps and weight-bearing distribution, as well as ground reaction force (GRF).	Kinetics, ground reaction force, pressure maps, gait analysis Limitations: setting up, placement	1000Hz, force, pressure
Electromyographic sensors (EMG)	Muscle contraction intensity.	Muscle activity (e.g., detecting contraction) Limitations: setting up, placement, batteries	EMG amplitude
Electromagnetic field sensors (EMF)	Use electromagnetic field for detecting and tracking objects based on Faraday's law of magnetic induction.	Motion sensors Limitations: setting up, placement, sensitivity to field variations	Up to 240Hz
Heart rate (HR)	Load/stress/activity level of the athlete. Heart rate variability (HRV).	Stress, load, exertion Limitations: setting up, batteries	Heart activity BPM, heart rate variability
INS/GNSS	Geolocation (latitude, longitude), sometimes velocity and elevation. Although these devices operate at lower frequency, with real-time kinematic (RTK) centimeter range precision is a possibility.	Geolocation, position Limitations: satellite, signal obstruction, batteries, frequency	10Hz

For biomechanics, sensors provide objective data that can be used for analysis and machine learning models. It's often impossible to capture micromovements and forces of a downhill racer, moving at 100 mph down Strife, a famed Austrian downhill run in Kitzbühel, but when equipped with the right sensors, we can capture data for biomechanical analysis, with data points available hundreds of times per second. In this book we discuss various IoT devices that can be used for sport data collection: computer vision cameras; image raging sensors, such as LIDARs; inertial measurement units (IMUs); attitude and heading reference systems (AHRS); inertial navigation systems (INS/GPS); pressure and electromyographic sensors; and others.

Deep Vision

The more observant you are, over time you begin to build up a real understanding.

—Frosty Hesson, "Chasing Mavericks"

Figure 5-3. Discobolus, 2nd century AD overlaid with pose estimation by neural net

Ancient art comes to us through millennia: from Paleolithic cave paintings dating back 40,000 years ago (Figure 5-4) to classic paintings and statues of the ancient Olympics in Greece. The ancient images can still be analyzed today by data scientists with deep learning methods; some of these methods are described in the chapters that follow on deep computer vision. Think of the longevity of information preserved in these sculptures and images!

Figure 5-4. Drawing a stickman was easier back in Stone Age, without deep learning!

In sports, the earliest and most reliable method of analysis and recording information is visual: from sketches and paintings, sculptures to photography and video, visual methods to analyze athletic performance have always been important to coaches. What's truly fascinating about visual data is that we can still use archive images of the athletes and sports of the past and apply machine learning methods to understand their technique. Figure 5-3 also reflects the importance of computer vision in applications for sports: the magical beauty of ancient Greek discus thrower, dated back to 2000 years, overlaid with the results of finding key body joints using a multilayer convolution neural net.

Today we can use RGB video at more than 60 frames per second at 4K resolution. In convolution neural nets, RGB planes are combined in the models to produce a single feature map. For example, Figure 5-5 shows a highly dynamic sport, tennis, yet today's machine learning models provide segmentation and pose estimation that may be sufficient for many applications. A word of caution though: publicly available datasets may still lack precision required, for example, for professional- and Olympic-level coaches. Accuracy of these models is as good as the data, so we need better datasets to satisfy the highest criteria of professional sports. We can add other sensors, for example, LIDARs, depth sensors, or multicamera installations, to get additional dimensions for modeling. Each technology has constraints, such as range, frequency, the need to have a direct line of sight, or calibration requirements.

Figure 5-5. Pose estimation of a tennis player

Deep learning methods are especially valuable for sports data scientists, kinesiologists, and biomechanists. 2D pose estimation (Chapter 7) and 3D body pose estimation methods (Chapter 8) may be used for AI-aided movement analysis, detecting sport-specific activities and movements. This is part of the job coaches do every day, but when athletes don't have a coach nearby, motion analysis is super-valuable! In Chapter 9, "Video Action Recognition," I will demonstrate how to apply deep learning methods to recognize activities for various sports, something that would be hard to do just a few years ago. In the past, this kind of analysis was done with a lot of manual work: sport scientists had to go over the video frame-by-frame, manually labeling every activity or marking a joint, but machine learning makes it much more efficient.

Since we are interested in biomechanical analysis, video is especially suitable for kinematics: it offers both *spatial* (movement in space) and *temporal* (movement in time) data. In fact, often, video is the only source of information available for kinesiology and biomechanical analysis.

Deep Vision Devices

We are all glorified motion sensors. Some things only become visible to us when they undergo change.

—Vera Nazarian, *The Perpetual Calendar of Inspiration*

There's a new class of devices for edge that starts taking shape, and it's coming from the need to do an onboard machine learning, for example, for robotics or self-driving cars. Most companies started to make boards with similar specs, fitting the profile of data science on the edge. The idea is to provide on-device computer vision capabilities in a small cost-efficient form factor. The list of new application-specific integrated circuits (ASICs) supporting tensor and matrix math required for deep learning at the edge keeps growing. Basically, the edge vision setups include a camera, a microSD card or storage to keep data while offline, Wi-Fi for online connectivity, a TPU or specialized circuit for accelerated processing, and a runtime for inference.

Basic Device

There's a wide range of camera devices: Raspberry Pi camera serves as an entry level to get started for a sport data scientist (Figure 5-6). At a cost less than a few dollars, it has several video modes including 1080p30 and 720p60 and image resolution of 8MP.

Figure 5-6. The simplest video capturing device equipped with a camera, Wi-Fi, and microSD storage at a cost of just a few dollars

In Chapter 3, "Data Scientist's Toolbox," Python was mentioned as one of the most popular languages for data science. Fortunately, the camera has an easy-to-use picamera Python module, which also works great with other computer vision libraries, including OpenCV. From the following code, it's easy to see that it takes just a few lines of code to capture a still image and record an H264 video:

```
import picamera

camera = picamera.PiCamera()
camera.capture('test-image.jpg')
camera.start_recording('test-video.h264')
time.sleep(5)
camera.stop_recording()
```

Even this simplest and least expensive setup can be used for field sports data science with offline or cloud-based processing. Raspberry Pi Zero includes microSD card, which can store gigabytes of data. If you have data connectivity, with Wi-Fi, you can do inferencing by building a cloud-based data science pipeline. To do inferencing onboard, however, we need an extra compute VPU to do any serious computation.

Edge Devices for Machine Learning

Most machine learning libraries require a coprocessing unit to accelerate computation. Historically, computers had at least one coprocessor, the GPU, primarily used to accelerate graphics. GPU has most of what deep learning computation needs: multicore parallel execution and memory.

In order to use the GPU from a generic code, there needs to be access to GPU's virtual instruction set. NVIDIA CUDA was created exactly to enable that: ability to execute code on the GPU and take advantage of GPU's parallel cores.

Figure 5-7. Jetson Nano, NVIDIA's entry point edge accelerated board

In this class of devices, NVIDIA Jetson Nano (Figure 5-7) costs less than a hundred dollars, is based on a quad-core Cortex A57 processor, and provides multiple CUDA cores for machine learning, with a rich set of GPU-optimized processing libraries. Most machine learning frameworks can leverage both CPU and GPU; the latter often requires more dependencies but offers a higher performance. For example, running sport activity body pose estimation on a CPU can get us 0.5–2 FPS; with GPU we can easily get 15–30 FPS.

Take a step further from the basic setup described earlier, by adding a vision optimized processing unit (VPU) to it, and you get a Google Vision Kit for less than a hundred dollars. With that you can do object detection and classification; the device includes a TensorFlow compiler to convert models for running on Vision Kit (Figure 5-8). The Vision Kit uses Raspberry Pi Zero W, which makes it a compact wireless, fully capable computer. Google Edge TPU (tensor processing unit; https://coral.withgoogle.com/) also provides a development board, a USB stick, and a camera.

Figure 5-8. Google Vision Kit, a low-cost but capable entry-level edge device

Intel Neural Compute Stick packs Movidius Myriad VPU and supports TensorFlow, Caffe, ONNX, PyTorch, and others via ONNX conversion (Figure 5-9).

Figure 5-9. Intel Movidius Neural Compute Stick

As this book is being published, several new products from Microsoft with AI Vision Dev Kit, TensorFlow Lite, for microcontrollers have been announced. Eventually, with developing powerful edge devices, capable of running video-related machine learning processing onboard, such as Movidius Myriad VPU, NVIDIA Jetson Nano, and Google Edge TPU, we will be able to use some of the methods described in this book on a device, without using an external compute resource.

For a deep dive in computer vision methods with practical Python and examples and notebooks, check chapters "Deep Computer Vision," "2D Body Pose Estimation," "3D Pose Estimation," and "Video Action Recognition" of this book. From object detection, classification to pose estimation in 2D and 3D, activity recognition, and skill analysis, computer vision offers many practical applications for movement analysis.

Inertial Movement Sensors

Did you know how quick your time was at the gate? No, I was by before I even thought about it.

—"Downhill Racer"

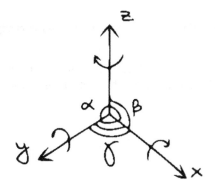

Inertial movement sensors take another approach at measuring movement: these devices combine accelerometers, gyroscopes, and in some cases magnetometers to measure acceleration, rotation, and direction. There're several reasons why these devices are absolutely fantastic for sports: for cameras you need a direct line of sight, while IMUs can be attached or integrated in clothing, do not require a line of sight, do not suffer from occlusion, and provide data at much higher frequency. What it means for a practical sport scientist or a coach: with an IMU you don't need to see the athlete, or point a camera at the field, IMUs will capture body movements and rotations. Trade-offs often include the need to calibrate devices and recharge multiple sensors. Also, IMUs do not measure position and velocity, although on many devices they are part of calculated metrics.

With IMUs we can easily apply mechanical terms to movement analysis (also see Chapter 2, "Physics of Sports"). Accelerometers readily provide acceleration, or change in velocity data:

$$acceleration = \frac{change\ in\ speed}{change\ in\ time}$$

For rotational movement, a gyroscope provides angular motion information:

$$angular\ velocity = \frac{change\ in\ angle}{change\ in\ time}$$

From Newton's second law, we know that acceleration helps linking kinematics and kinetics, the acceleration of the body is directly proportional to the magnitude of the unbalanced force, and the acceleration of the body is inversely proportional to the mass of the body:

$$force = acceleration * mass$$

IMUs typically provide six or nine "degrees of freedom" (6DOF or 9DOF), which is measurement of movement with 6 or 9 pieces of data: vectors of acceleration (x,y,z), angular rotation (x,y,z), and magnetic force (x,y,z). This data is provided at high frequency of up to 200Hz and higher, depending on the sensor.

Basic IMU

IMU devices have come a long way from the IMU pictured in Figure 5-10, which shows a drawing of an Apollo 5 IMU, used for navigation in space back in 1968, mainly thanks to MEMS technology that fits an integrated IMU, containing accelerometer, a gyro, and a magnetometer all in a tiny 4 square millimeter component. There're several integrated edge devices with embedded IMUs, including the one myself and my partners built, pictured in Figure 5-11. These sensors are good for kinematic analysis of multiple sports, including basketball, tennis, and skiing, if you are willing to accept certain limitations, such as the need to charge and calibrate.

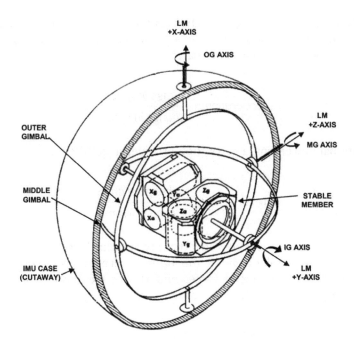

Figure 5-10. Apollo 5 IMU was so much bigger than today MEMS IMUs! (Source NASA)

Figure 5-11. A smartwatch size prototype of an integrated edge IoT device with an IMU

Take Olympic downhill, for example: the famed Kitzbühel Streif has a length of more than 3 kilometers, with a vertical drop of almost a kilometer, maximum angle of 85 degrees, and many curves, blocking cameras. Skiers reach speeds of 145 km/h: it is nearly impossible to provide an accurate continuous line-of-sight camera coverage there, with a precision required to monitor every joint of the skier! But when equipped with IMUs (better even a suit that has IMUs built-in, like an Xsens, for example), we can record data from multiple sensors, reading every joint of the body at a rate of 200Hz or more, regardless of the terrain complexity! The trick? Placing all those sensors on the body, calibrating them, and saving the data. However, data scientists should be fascinated by the quality of data coming from these devices, especially for good high-precision datasets! Note that these sensors can measure relative movement of the body limbs; absolute position can only be available by adding a device capable of tracking location. Sensors can be integrated in sports equipment, see for example this sensor prototype I built for surfing (Figure 5-12).

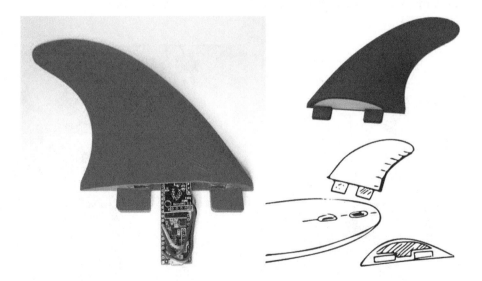

Figure 5-12. Surfing fin sensor based on IMU to estimate surfboard movements

There're many ways to apply these devices for sport data science. One of the advantages is a very high frequency of measurement; any mobile phone these days is equipped with an IMU: a linear accelerometer typically reads at 100–200Hz, typically it's higher than what cameras can provide, and a gyroscope returns a 3D vector, measuring angular rate at 200–2000 dps (degrees per second). Magnetometers provide data in gauss range 5–16 gauss. Together, IMUs consist of solid-state or microelectromechanical (MEMS) gyroscopes, accelerometers, and magnetometers.

Attitude and Heading Reference System

The key addition to attitude and heading reference system (AHRS) to the set of three IMU sensors is the system that computes *attitude* and *heading*. Such a system typically employs a filter to combine the output of sensors. In small systems we may use complementary, Mahony, Madgwick, or Kalman, filters for data fusion. These filters take raw accelerometer and gyro data and calculate an orientation quaternion or Euler's angles.

Attitude and heading reference systems have many applications. During my work on sensors, I met Cyrus Hostetler, a brilliant javelin thrower and an Olympian on the US Olympic Team, who came up with an idea of attaching an IMU sensor to a javelin to provide measurements for training. We quickly put a prototype together (see Figure 5-13). After some tweaking with the hardware and printing a 3D case that fits the javelin, we did some interesting tests, measuring javelin throwing angle, rotation, and in-flight data.

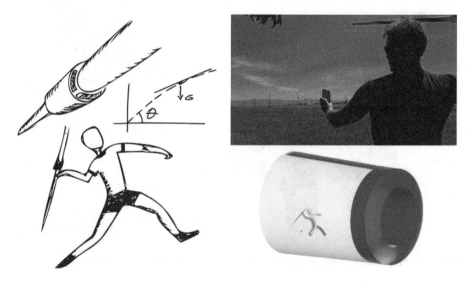

Figure 5-13. An IMU prototype to analyze javelin throw

Inertial and Navigation Systems

In order to have a complete description of the motion, we must specify how the body alters its position with time; i.e. for every point on the trajectory it must be stated at what time the body is situated there.

—Albert Einstein, Relativity

Global navigation satellite systems, or GNSS (the term that includes GPS, GLONASS, Galileo, BeiDou, and other regional systems), have become a household name for getting from point A to point B. We rely on these systems, with receivers built into many devices, including our phones, to get directions. My earliest venture into the world of navigation and positioning was by building Active Fitness and Winter Sports apps: both rely on location APIs in the phone to track running, walking, skiing, and snowboarding along with 50+ other activities for several million users. With inexpensive boards like a RaspberryPi, making custom sensors that also include a navigation component becomes feasible (Figure 5-14).

Figure 5-14. A sensor based on Raspberry Pi Zero with a GPS+IMU and wireless charging

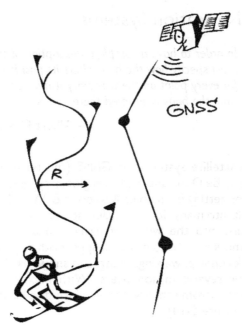

Figure 5-15. GNSS precision and actual turns of the skier: when knowing location is not enough!

While these systems provide geopositioning data (latitude, longitude, sometimes elevation and calculated speed), the best frequency at which these devices can provide data is about 1–10Hz. Compare that with 100Hz for an IMU! In downhill, skiers often reach speeds of 90 mph (44 meters per second) or more; that means that with GNSS rate of 1–5 times per second, we will miss most skier's turns information that is crucial to the coach and the athlete. In reality, our precision is in the range of 5–10 meters, not sufficient to get a smooth curve of a single skier's turn (Figure 5-15).

For sports it means that we can get an approximate course of the skier on the slope, but not a smooth curve of a turn with GNSS alone!

The satellite signal for GNSS is not always available: even while skiing in the mountains, satellite signal can be easily blocked by rocks and trees. To overcome that, inertial navigation systems (INS) continuously calculate position by *dead reckoning* and using IMUs (accelerometer, gyro, and magnetometer) to estimate a calculated position. From physics we know that acceleration is the first derivative of velocity by time and the second derivative of displacement by time:

$$a = \frac{dv}{dt} = \frac{d^2s}{dt^2}$$

Since IMUs provide acceleration data, in order to estimate at what speed and how much we moved relative to dead reckoning, we need to integrate acceleration to get speed and then integrate speed to get displacement:

$$v = \int a\,dt, \quad s = \int v\,dt$$

Of course, integration accumulates an error, or drift. With time, drift can be very significant; getting a new reading from the geolocation unit helps correcting it, but if the signal is lost for a significant amount of time? More advanced INS systems that require centimeter-level precision use real-time kinematic (RTK), a complex method involving an additional baseline or a base station.

Range Imaging Sensors

Range imaging provides additional data, showing the distance to points in the scene. The basic principle on which LIDARs operate can be described by this formula:

$$distance = \frac{\left(speed\ of\ light * time\ of\ flight\right)}{2}$$

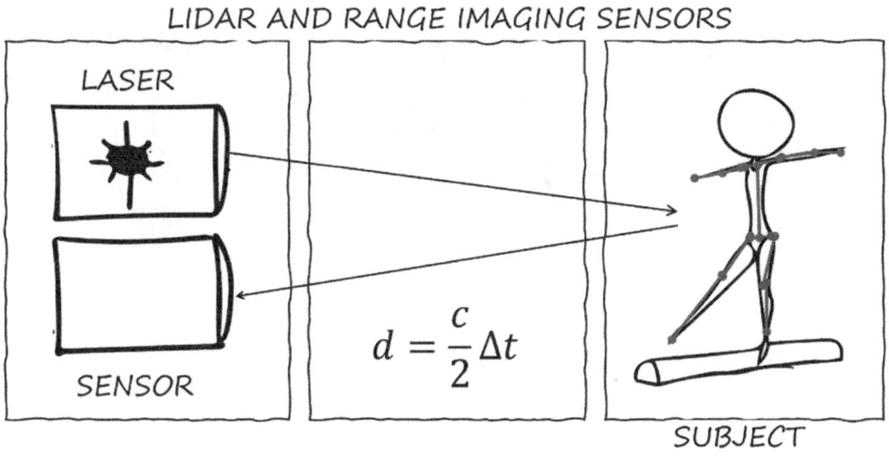

LIDAR AND RANGE IMAGING SENSORS

LASER

$$d = \frac{c}{2}\Delta t$$

SENSOR

SUBJECT

Figure 5-16. LIDAR sensors measure distance with reflected laser light

Since the speed of light is known and time is measured from reflection, the sensor gets a very accurate distance to the object, something a regular camera cannot provide.

There're many ways to get depth data: stereo and sheet triangulation, structured light (used in 3D scanners), time of flight (LIDAR), interferometry (InSAR, terrestrial synthetic aperture radar), and coded aperture (method that may apply principal component analysis, or PCA, and is used in iPhone X, for example). In sports we often need to focus on the subject athlete at high speeds in various light environment, so time-of-flight methods such as LIDAR are used successfully (Figure 5-16).

Recently, for the first time in history of gymnastics sport, an AI system that uses a combined computer vision and LIDAR is used to judge world-level gymnastics competitions. The International Gymnastics Federation (FIG) together with Fujitsu announced in 2018 that they are working on the system combining LIDAR and computer vision to implement a fair judging system for gymnastics.

Pressure Sensors

Have you noticed that whatever sport you're trying to learn, some earnest person is always telling you to keep your knees bent?

—Dave Barry

PRESSURE SENSOR

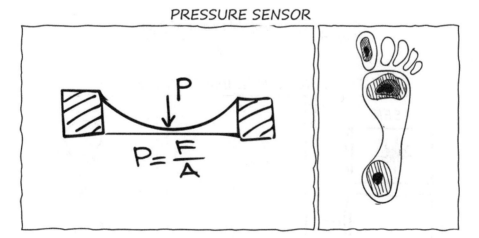

Figure 5-17. Pressure sensors: measuring force applied over area

Pressure sensors are widely used in sport movement analysis. Think of athletic stance: our stance in many sports defines how we move. But the subtle position of our feet (Figure 5-17), pressure areas, for example, on top of a surfboard, or inside ski boots, is not easy to see with vision or from inertial sensors data. Your best bet if you need foot pressure is pressure sensors. Fortunately, some pressure sensors are really inexpensive.

The first type of devices I cover in this book is measuring contact pressure and may involve capacitive sensors or piezoelectric sensors. Both of these sensors are great for sports applications, because they can measure pressure distribution or contact pressure essential for some areas of movement analysis. Capacitive sensors use change in capacitance to measure pressure. Two conducting plates are separated by a nonconductive material. When pressure is applied, the plates get close and the capacitance changes. This can be measured and hence reflects the pressure value. Piezoresistive sensors change a voltage when squashed. For a membrane with a known area that deforms under pressure force, pressure can be calculated as:

$$pressure = \frac{force}{area}$$

Another pressure sensor type is barometric: these are part of most robotics today and provide a high-precision altitude measurement. These are best when integrated with IMUs/AHRS/INS systems and are covered in the sections related to motion and position sensors.

EMG Sensors

Electromyographic (EMG) sensors measure muscle activity. This type of sensor is particularly useful for measurements of muscle contraction and provides data about muscle activity (Figure 5-18).

EMG SENSOR

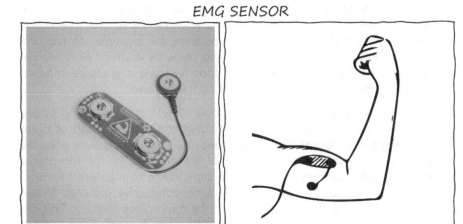

Figure 5-18. EMG sensor measuring muscle contraction of the arm

Heart Rate

 Heart rate (HR) sensors are used to measure load and exertion. It was one of the most popular sensors at the turn of the millennium, and it remains today one of the most objective ways to measure the workload. Most smartwatches these days are equipped with HR monitor of some sort, so if you have access to heart rate data, you can correlate that information with your movement analysis.

The frequency of data coming out of heart rate monitors is low, but it may provide useful insights into injury analysis, stress, and factors related to risk, such as fatigue. Another type of measurements for heart rate is heart rate variability (HRV), which is variation of time interval between heart beats. This measurement provides interesting insights into nervous system balance and reactions and serves as a good indicator of the nervous system health.

Summary

In this chapter we looked at various types of sensors a data scientist can use for health and fitness analysis. From biomechanics, we know principles of analytical approach to movement: kinetics and kinematics. Sensors are data collection devices: applying the right sensor to the problem is important, as we've seen in this chapter. As data scientists, we need to know the type of data that comes out of devices, frequency, accuracy, and possible applications to movement analysis. In the next chapters, we'll discuss machine learning methods and models available to a sport data scientist today. For additional materials and a video course, supplementing this book, check my Web site ActiveFitness.AI http://activefitness.ai.

Applying Machine Learning

All you need is lots and lots of data and lots of information about what the right answer is, and you'll be able to train a big neural net to do what you want.

—Geoffrey Hinton

Deep Computer Vision

Neuroscience and Deep Learning

Neural networks that were trained to discriminate between different kinds of images have quite a bit of the information needed to generate images too.

—Alexander Mordvintsev, Author of Deep Dream

In Chapter 4, "Neural Networks," we explored similarities between neurons and mathematical abstractions, present in artificial neural nets (ANNs). While some deep learning elements – forward propagation, activation functions, and weight transfer – naturally fit into both mathematical and biological models, some ANN algorithms, such as backpropagation, are harder to explain biologically. Brain may use other optimization mechanisms for training its neurons.

K. Ashley, *Applied Machine Learning for Health and Fitness*,
https://doi.org/10.1007/978-1-4842-5772-2_6

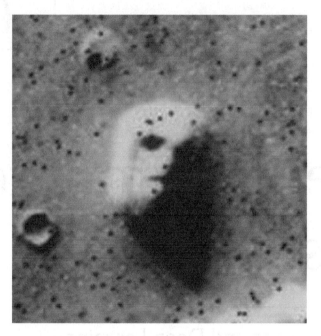

Figure 6-1. "Face on Mars" is an example of pareidolia (NASA)

Despite differences, computation results of artificial neural nets and brain are strikingly similar, including perception of optical illusions. *Pareidolia* is a well-known phenomenon: most of us will see a human face on this Martian rock formation picture taken by Viking orbiter (Figure 6-1); however from a different angle, it looks just like another rock. Our tendency to see familiar shapes and even facial expressions often requires only a few pen strokes; the physical nature of this tendency is also supported by neuroscience. Conveniently, if we apply machine methods for detecting features (such as a smile in Figure 6-2, which contains three sticks and one circle!), we may get the same results as the human brain. And this face is not even smiling, but it tells me where the smile is supposed to be! The method I use here is well known for facial recognition; it's called Haar Cascade classifier and is included in OpenCV. Let's see if we can use it to detect a smile (Figure 6-3)!

Figure 6-2. Our brain eagerly tells us that this is a human face and tries to identify its expression

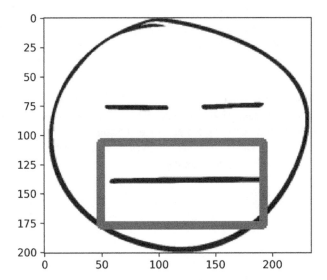

Figure 6-3. You can trick human brain and machine learning into believing three sticks and one circle is about to smile!

```
image = cv2.imread('media/face.png')
image = cv2.cvtColor(image, cv2.COLOR_BGR2RGB)
plt.imshow(image)
plt.show()
cascade = cv2.CascadeClassifier(cv2.data.haarcascades+'haarcascade_smile.xml')
features = cascade.detectMultiScale(image)
print(features)
```

```
for (x, y, w, h) in features:
    img = image.copy()
    cv2.rectangle(img, (x, y), (x+w, y+h), (0, 255, 0), 6)
    plt.imshow(img)
    plt.show()
```

The idea that networks trained to recognize various types of images can also generate images prompted Alexander Mordvintsev, an engineer from Google, to create Deep Dream, a model that generates dreamlike bizarre images based on what it learns.

Computer Vision in Health and Fitness

Of all the methods to analyze health, fitness, and biomechanical information, computer vision stands out: most information we have collected is visual, from the earliest stone cave drawings of physical activities to modern days with nonstop sport events broadcasting. Not surprisingly, a sport data scientist toolkit has grown quite extensive: from general-purpose computer vision libraries, integrating computer vision and deep learning components, such as OpenCV, scikit-learn, PyTorch, TensorFlow, and other machine learning libraries.

We also have a broad range of activities analyzed with machine learning in publicly available datasets than we had just a decade ago (Figure 6-4). It's important to note that some sport activities require a much higher frequency than the video to provide adequate analytics, and devices such as sensors mentioned in this book may still be required.

Figure 6-4. Computer vision in sports has public datasets and models covering a broad range of activities

Loading Visual Datasets

In data science 90% of time spent finding data and 10% of time spent complaining about collecting the data.

—Old data science joke

The first question any data scientist asks is: "Can you give me the data?" Deep learning output is only as good as your training data. In the last few years, datasets evolution made possible many amazing data science applications. We already used some of these datasets (e.g., COCO), to help us make first steps in sports science, estimating positions of athletes in 2D and 3D and recognizing human activities. Working with multiple images and models in high-level frameworks such as PyTorch is simplified and typically involves loading and transforming data, training the model, or using an existing model. For loading, ImageFolder and DatasetFolder are two generic objects that help loading images and data for training and scoring: for example, ImageFolder makes it easy to load image data, stored in folder structure that matches root/ class/*.png and apply transformations:

```
t = transforms.Compose([transforms.Resize(256),
        transforms.CenterCrop(224),
        transforms.ToTensor(),
        transforms.Normalize(
        mean = [0.485, 0.456, 0.406],
        std = [0.229, 0.224, 0.225])])
train_data = datasets.ImageFolder(f, transform=t)
```

We need both test and training data, so in my load method, I load images from activities folder and PyTorch automatically labels them with classes:

```
trainloader, testloader = load(f, .2)
print(trainloader.dataset.classes)
images, labels = next(iter(trainloader))
grid = torchvision.utils.make_grid(images)
plt.imshow(grid.permute(1,2,0))
['surfing', 'tennis']
```

In the projects that follow, we will use this dataset to train the model and classify sport activities. But for now, let's discuss the heart of machine learning, our models!

Model Zoo

A visit to a city can't be complete without a stop by a local zoo. When I moved to Bay Area, my checklist included visiting Golden Gate Park and checking out the famous San Francisco Zoo. But it turned out, California local wildlife frequented my house anyway, without bothering with property boundaries. I'm also fortunate to live by a running course in golden California hills, established in 1971, where I can observe local animals every day: deer, rabbits, coyotes, and occasionally, bobcats.

In data science we have model zoos (Figure 6-5): collections of pretrained models that can be used for various tasks. Some frameworks, like PyTorch, include built-in methods to fetch models and start using them quickly. Some of these models are pretrained for vision, audio, and text tasks. Conveniently, in PyTorch, they are called torchvision, torchaudio, and torchtext. Some of them are particularly interesting for us because they work great in sport movement analysis.

Figure 6-5. Model zoo

Applying Models

Torchvision models include image classification (AlexNet, VGG, ResNet, etc.), semantic segmentation (FCN ResNet, DeepLabV3), object detection (Faster R-CNN, Mask R-CNN), keypoint detection (Keypoint R-CNN ResNet-50 FPN), video classification (ResNet 3D), and so on. You can also create a custom model and add it to the zoo. Models come pretrained (weights included), or blank with random weights.

Figure 6-6. Quiet, please: machine learning!

Data science applications in sports often require predicting the most efficient way to make movements, and that requires both deep domain knowledge and models. When should we use rules and when machine learning? Machine learning is not the panacea for all problems: oftentimes basic statistical methods or rule-based programming applies best. The rule of thumb is when the problem can be described by simple rules, we can use rules. When rules are hard or impossible to create deterministically, we use machine learning (Figure 6-6).

Now, we are ready to apply our knowledge to do one of the most fundamental tasks in machine learning: classifying our data!

Classification

PROJECT 6-1: CLASSIFYING ACTIVITY TYPE

In this project we will classify images from different sporting activities, using ResNet:

```
import torch
import torch.nn as nn
import torch.optim as optim
from torch.optim import lr_scheduler
```

```
import torchvision
from torchvision import datasets, models, transforms

device = torch.device("cuda" if torch.cuda.is_available() else "cpu")
model = models.resnet18(pretrained=True)
```

In the previous sections, we loaded a small set of data, containing 100 images for 2 classes: surfing and tennis. This is not much data to train our model, but fortunately we can use a technique called transfer learning and leverage ImageNet trained on millions of images. Remember that PyTorch auto-calculates gradients, so for transfer learning, we need to freeze gradients by setting requires_grad to False to prevent PyTorch to recalculate gradients during backpropagation (see Chapter 4 for neural networks training and explanation of how backpropagation works):

```
for param in model_conv.parameters():
        param.requires_grad = False
```

Next, we use a bit of magic here: in data science speak, it's called using ConvNet as fixed feature extractor. It means we'll remove the last fully connected layer that comes with the pretrained model and replace it with only the linear classifier for two classes we need, surfing and tennis:

```
features = model.fc.in_features
model.fc = nn.Linear(features, len(labels))
model = model.to(device)
criterion = nn.CrossEntropyLoss()
optimizer = optim.SGD(model.parameters(), lr=0.001, momentum=0.9)
scheduler = lr_scheduler.StepLR(optimizer, step_size=7, gamma=0.1)
```

Now, we can train the model and save it for the future:

```
model = train_model(epochs=3)
torch.save(model, 'activity_classifier_model.pth')
```

Let's see the plot (Figure 6-7) of how well our model converges, by adding a line plot for the train and test loss values:

```
plt.plot(train_losses, label='Training loss')
plt.plot(test_losses, label='Validation loss')
plt.legend(frameon=False)
plt.show()
```

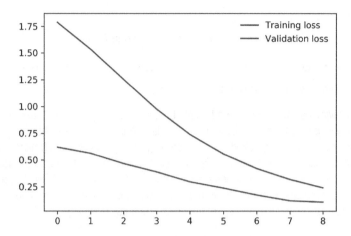

Figure 6-7. Model convergence for activity classification

Finally, we can reload the model and give it a testing set of images to see how well it does, and if the model can really tell one activity from another.

```
data_dir = 'data/activities/'
test_transforms = transforms.Compose([transforms.Resize(224),
                                      transforms.ToTensor(),])
device = torch.device("cuda" if torch.cuda.is_available() else "cpu")
model=torch.load('activity_classifier_model.pth')
model.eval()
```

The result (Figure 6-8) looks pretty good; you can see that images of different activities have been identified by the model correctly.

Figure 6-8. Test images classification for sport activities with the trained model

Detection

**PROJECT 6-2: DETECTION: APPLYING FASTER R-CNN
MODEL TO DETECT A SURFER AND A SURFBOARD**

Faster R-CNN model included with PyTorch is trained to predict multiple classes from COCO categories, including sports equipment, such as skis, snowboard, surfboard, tennis racket, baseball glove, and skateboard, which makes it pretty useful in sports. We will use Faster R-CNN model to demonstrate how to identify a surfboard on the image of a surfer (Figure 6-9).

Figure 6-9. Surfer on the wave

Let's start by loading a model in PyTorch; pretrained flag set to True instructs PyTorch to use a pretrained model:

```
model = torchvision.models.detection.fasterrcnn_resnet50_fpn(pretrained=True)
model.eval()
```

Thanks to torchvision, if we don't have the model, it gets downloaded automatically into .cache:

```
Downloading:    "https://download.pytorch.org/models/fasterrcnn_resnet50_
fpn_coco-258fb6c6.pth" to .cache\torch\checkpoints\fasterrcnn_resnet50_
fpn_coco-258fb6c6.pth
100.0%
```

Next, I created `filter_detections` method that runs predictions, based on our model, and filters predictions using a threshold of 0.7 confidence (we are not interested in surfboards with confidence level less than 70%)!

```python
def filter_detections(img, model, threshold=.7):
    with torch.no_grad():
        img_t = T.ToTensor()(img)
        img_t = img_t.unsqueeze(0)
        if next(model.parameters()).is_cuda:
            img_t = img_t.pin_memory().cuda(non_blocking=True)
        pred = model(img_t)[0]
    boxes = pred['boxes']
    box_scores = pred['scores']
    labels = pred['labels']
    idxs = [i for (i,s) in enumerate(box_scores) if s > threshold]
    res = [(COCO_CLASSES[labels[i].cpu().tolist()],boxes[i].cpu().numpy())
    for i in idxs]
    return res

detections = filter_detections(img, model)
print(detections)
```

The result is bounding rectangles for detected instances of a person and a surfboard!

```
[('person',
  array([220.0556 ,  75.23606, 305.56512, 191.46136], dtype=float32)),
 ('surfboard',
  array([232.77135, 137.87663, 365.316  , 199.95457], dtype=float32))]
```

Finally, let's draw resulting bounding boxes for the surfer and the surfboard (Figure 6-10):

```python
from PIL import Image
from PIL import ImageDraw, ImageFont

def show_detections(img, detections):
    colors = {"surfboard":"lime","person":"red"}
    draw = ImageDraw.Draw(img)
    font = ImageFont.truetype("arial",16)
```

```
for d in detections:
    name = d[0]
    box = d[1]
    color = "red"
    if name in colors:
        color = colors[name]
    draw.rectangle(((box[0], box[1]), (box[2], box[3])), outline=color,
    width=8)
    draw.text((box[0], box[1]-24), name, fill=color, font=font)
return img

img_detections = show_detections(img,d)
plt.imshow(img_detections);
plt.show()
```

Figure 6-10. Detection of the surfer and surfboard with Faster R-CNN model

Segmentation

PROJECT 6-3: APPLYING RESNET MODEL FOR SEMANTIC SEGMENTATION

In the previous project, we detected an instance of an object. Now, we'll play with another model to isolate a segment on the image. This is called semantic segmentation (Figure 6-12), a method widely used in deep learning and computer vision. To illustrate our very practical use case: selecting an athlete on the image, I'll be using a pretrained FCN ResNet 101 model:

```
from torchvision import models
```

```
fcn = models.segmentation.fcn_resnet101(pretrained=True).eval()
```

Notice `pretrained=True` flag in the model constructor: I could also choose a blank model without weights, and then I would simply specify a constructor without arguments. If you run this line of code for the first time, and the model hasn't been cached yet, torchvision will automatically download and store the model in the cache. Pretty cool, huh? For a data scientist, having to deal with multiple models, that's a relief. To list models available from torchvision, for example, run this command:

```
torch.hub.list('pytorch/vision')
```

```
['alexnet',
 'deeplabv3_resnet101',
 'densenet121',
 'densenet161',
 'densenet169', ...
```

Without a transformation, even humans looking at some images may be confused, let alone our models! Look at Figure 6-11, for example, and see if you can tell a bunny from a duck without a proper rotation! Pretrained models are often expected to have data normalized in the same way: for images in pretrained torchvision models, this involves transforming them to three channels (RGB) * h * w, where h and w are at least 224 pixels in [0,1] range and normalizing using specific mean and standard deviation. That's actually pretty cool that models we use can perform tasks using such tiny images!

Figure 6-11. Bunny or duck? Transforming data before processing with the model

To transform images in a form that's best suited for pretrained models, we use a composition of several transformations:

```
import torchvision.transforms

normalize = transforms.Compose([transforms.Resize(256),
                transforms.CenterCrop(224),
                transforms.ToTensor(),
                transforms.Normalize(mean = [0.485, 0.456, 0.406],
                        std = [0.229, 0.224, 0.225])])
img_tensor = normalize(img)
```

In the preceding sequence of transformations, the image is resized and is cropped to 224 points with Resize and CenterCrop transforms; then the PIL image in the range of [0,255] is converted to float tensor (CxHxW) in the range of [0,1] with ToTensor() transform, and then normalized. Now, this image is ready for our model:

```
out = fcn(img_tensor.unsqueeze(0))['out']
out_model = torch.argmax(out.squeeze(), dim=0).detach().cpu().numpy()
print (out_model.shape)
```

and display the result:rgb = show_segment(out_model)

plt.imshow(rgb); plt.show()

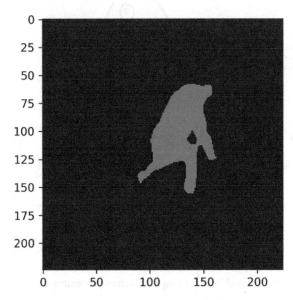

Figure 6-12. Segmentation result with the athlete body highlighted

To help models train on larger datasets, most machine learning frameworks include methods for synthetic data augmentation, which may involve rotating or transforming images automatically.

Human Body Keypoints Detection

PROJECT 6-4: APPLYING RESNET MODEL TO DETECT HUMAN BODY KEYPOINTS

In the previous example, we did semantic segmentation, using a pretrained FCN model. In this project, we'll dive deeper into another practical example of detecting an athlete's body keypoints. We'll use Keypoints R-CNN ResNet FPN model trained on COCO dataset classes for surfer's body detection.

```
from torchvision import models

kprcnn = models.detection.keypointrcnn_resnet50_fpn(pretrained=True)

kprcnn.eval()
```

We will use the same image tensor transformation, from the previous example, but this time, we'll pass it to Keypoints R-CNN model for human body keypoint detection:

```
detection = kprcnn(img_tensor)[0]
detection
```

Now, the result looks different from detections in the previous examples; let's convert it into JSON to make easier to understand:

```
keypoint_detections = get_preds(img)
json_keypoints = to_json(keypoint_detections)
json_keypoints
```

```
'[{"nose": [924, 420, 1], "left_eye": [923, 408, 1], "right_eye": [920,
405, 1], "left_ear": [884, 404, 1], "right_ear": [895, 405, 1], "left_
shoulder": [803, 415, 1], "right_shoulder": [910, 478, 1], "left_elbow":
[789, 408, 1], "right_elbow": [924, 562, 1], "left_wrist": [954, 636, 1],
"right_wrist": [951, 636, 1], "left_hip": [723, 580, 1], "right_hip":
[751, 610, 1], "left_knee": [834, 570, 1], "right_knee": [887, 613, 1],
"left_ankle": [867, 753, 1], "right_ankle": [865, 759, 1]}]'
```

Finally, we can visualize vectors connecting keypoints to create a stickman pose estimated by the model (Figure 6-13):

```
def drawline(draw, z, f, t, c):
    if f in z and t in z:
        draw.line((z[f][0], z[f][1], z[t][0], z[t][1]), fill=tuple(c),
        width=12)

def gen_colors():
    arr = plt.cm.tab20(np.linspace(0, 1, 20))
    arr=arr*255
    return arr.astype(int)

def draw_stickman(img):
    draw = ImageDraw.Draw(img)
    data = json.loads(json_str)
    z = data[0]
    colors = gen_colors()
    drawline(draw, z, 'left_hip', 'nose', colors[0])
    drawline(draw, z, 'left_hip', 'left_knee', colors[1])
```

```
        drawline(draw, z, 'left_knee', 'left_ankle', colors[2])
        drawline(draw, z, 'left_shoulder', 'nose', colors[3])
        drawline(draw, z, 'left_shoulder', 'left_elbow', colors[4])
        drawline(draw, z, 'left_elbow', 'left_wrist', colors[5])
        drawline(draw, z, 'right_hip', 'nose', colors[6])
        drawline(draw, z, 'right_hip', 'right_knee', colors[7])
        drawline(draw, z, 'right_knee', 'right_ankle', colors[8])
        drawline(draw, z, 'right_shoulder', 'nose', colors[9])
        drawline(draw, z, 'right_shoulder', 'right_elbow', colors[10])
        drawline(draw, z, 'right_elbow', 'right_wrist', colors[11])
    plt.imshow(img)

draw_stickman(img)
```

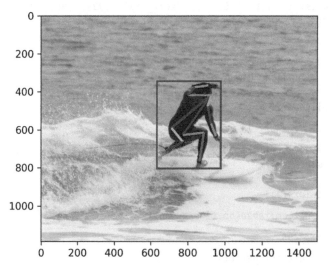

Figure 6-13. Body keypoint detection with Keypoints R-CNN

What we've done here is very useful for movement analysis in sports. We will use these results in the following chapters, discussing computer vision applications in sports.

Summary

In this chapter we covered some deep computer vision tasks, such as detection, classification, and object segmentation, using practical examples in sports. We looked at parallels between neuroscience, the way the human brain analyzes images, and neural networks. In this chapter I included several practical projects, demonstrating how to use some of the models, like R-CNN and ResNet; I used some of the most popular data science libraries, like PyTorch, that include useful models for deep vision tasks we identified earlier. In the next chapters, we'll dive deeper into methods for 2D and 3D body pose estimation, video action recognition, and other computer vision methods that may be useful for applications in sports, health, and fitness.

2D Body Pose Estimation

Nothing happens until something moves.

—Albert Einstein

Background

Pose estimation is the task of determining a human pose from an image or a sequence of images. It also serves as a first step for mapping the pose in 3D, which is covered in the next chapter. For biomechanics, the method provides a lot of useful kinematic information and has a multitude of applications. In the past, scientists often needed to go through images or video frames manually, mapping joints and limbs. With machine learning we now have models that do the work automatically, in near real-time (Figure 7-1). Pose estimation is used in gaming, film industry, health, robotics, and sports and recently more widely in other areas, such as retail. In the last few years, deep learning models in computer vision have experienced a breakthrough: methods to predict human poses are getting faster and more precise. For sports, pose estimation helps identifying key points in the human body, for performance analysis or injury prevention, which is essential for professional athletes and coaching.

© Kevin Ashley 2020
K. Ashley, *Applied Machine Learning for Health and Fitness*,
https://doi.org/10.1007/978-1-4842-5772-2_7

Figure 7-1. Pose estimation of a surfer in action

Methods

For a successful technology, reality must take precedence over public relations, for nature cannot be fooled.

—Richard Feynman

There're two approaches for 2D pose estimation that most models use: *top-down* and *bottom-up*. Bottom-up approach is the fastest; it focuses on detecting keypoints first and then groups them into poses. The top-down approach detects people in the scene and then applies a single-person keypoints detection. For fast action sports, it may be slower, because it depends on the number of people in the scene, and the risk of failure is higher, because of the early commitment to detect individuals, rather than focusing on joints and individual movements.

▨ **2D pose estimation approaches** Bottom-up approach recognizes joints first and then combines them into a pose. Top-down pose estimation works by detecting groups of subjects and then applies a single-person detector; for multi-person scene, its performance depends on the number of subjects.

Before diving into practical projects, let's go over to the basic concepts of pose estimation: methods, datasets and data points, benchmarks, and tools a sport data scientist can use.

Most human pose estimation methods use multistage recurrent neural networks (RNNs). The main reason being that video processing, which is the main source of sport analysis, unlike single images, benefits from sequential weight sharing between frames. So, for video, recurrent neural networks that can learn long sequences, such as LSTM (long short-term memory), between video frames are widely used in many greatly performing models. Some models also use temporal information provided by classic computer vision methods, such as optical flow to decrease flickering and improve prediction smoothness by tracking joints over time from frame to frame.

There're several deep learning models that work well for pose estimation; some of them are listed as follows:

- Mask R-CNN, a two-stage method that works by generating proposals (areas of the image) in the first stage and then creates masks or bounding boxes around areas.

- Stacked Hourglass Networks use a method of consecutive top-down, bottom-up steps in conjunction with intermediate supervision for better capturing of spatial relationships associated with the body.

- Cascade Pyramid Networks method deals with keypoint occlusion problems and targets keypoint localization for the human body.

- Part Affinity Field method (used, e.g., in OpenPose), a bottom-up system with real-time characteristics.

Datasets

For 2D pose estimation, several widely used datasets are available. Typically, pose estimation models are trained with some of these datasets, providing good accuracy:

- COCO (Common Objects in Context) dataset is based on 330K images and a quarter of a million people with keypoints. Microsoft and many other partners contributed to that dataset. The goal is placing the object recognition in the broader context of a scene understanding.

- MPII includes 25K images with over 40K people and 410 activities: images for this dataset are extracted from YouTube.

- LSP (Leeds Sports Pose Dataset) contains 2K pose-annotated images taken from Flickr and includes 14 joints detected.

- FLIC (Frames Labeled in Cinema) has about 5K images from Hollywood movies and running a person detector on certain frames.

■ **Note** Some datasets available online may be automatically generated (e.g., from games) or other sources and may not contain precise athletic movements. Recording high-precision custom datasets with labeled athletic data often requires additional sensors.

For pose estimation, we expect the data output as points of the human body. Body part mapping may be different, depending on what parts of the body you are interested in as a data scientist, and it doesn't always match precise human joints. For example, a typical data output using a COCO dataset may include multiple points and skeletal body parts, as illustrated in Figure 7-2.

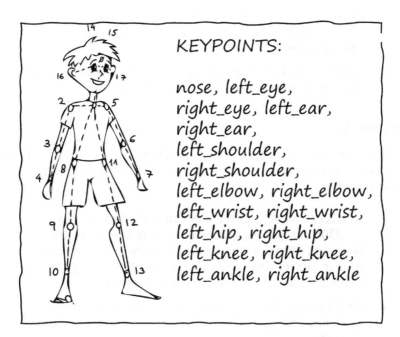

Figure 7-2. Keypoints from COCO dataset

Let's step back for a moment to understand how the concept of humanoid models and body parts applies to sport movement analysis. When coaches talk about movement analysis, they typically split the body of a moving athlete into areas of interest. In biomechanics, for example, the human body is typically analyzed in frontal, sagittal, and transverse planes.

When I was an aspiring ski instructor, the first thing I learned for movement analysis was separating movements of the "jacket" (upper body) from "pants" (lower body). It turns out, this simple categorization of looking at the upper and lower body of a moving athlete provides enough first glance information for a ski instructor to detect a skill level. Most beginners would typically rotate upper body to turn, while a good skier would maintain a quiet upper body and mostly have motion initiated from the lower body: ankles and knees, for efficient, balanced turns.

Most neural nets work similarly to human coaches: from detecting areas of interest to segmentation and classification. In our practical projects further in this chapter, we will take that a step further to analyze real sport activities.

Benchmarks

Benchmark scores for body pose estimation across many methods have improved a lot over the last few years. That tells us that computer vision is making gigantic strides toward understanding of human motion and activities. In just one year, for pose estimation, COCO mean average precision metric has increased from 60% to 72.1%, and another benchmark for MPII dataset jumped from 80% to almost 90%! Even an Uber driver after an AI symposium in Sunnyvale asked me about Skynet from the old Terminator movie, and how close we are to what's predicted in the movie back in 1984. I'm not sure about Skynet, but MobileNet, PoseNet, and many other convolution neural nets are doing very well!

Evaluation metrics frequently used in pose estimation are percentage of correct parts, percentage of correct keypoints, percentage of detected joints, and mean per joint position error. These metrics are frequently mentioned and tracked to estimate performance of pose estimation models.

Tools

For a sport data scientist, there's a wide selection of tools available for pose estimation. Practical projects in this chapter use various tools, to get you familiarized with a wide selection of technologies, rather than focusing on a single toolset. Most tools use Python, since the language became the lingua franca of data science, so we'll continue this tradition and will use scikit-learn, OpenCV, PyTorch, and TensorFlow for most of our projects.

It's worth mentioning that most deep learning frameworks come packaged with GPU-accelerated libraries. I highly recommend spending a little bit of time to make sure that an accelerated version of these libraries works in your environment! Deep learning using videos is resource intensive, and the difference between 0.2 frames per second and 30 frames per second is significant if you are a sport data scientist. Most Python packages have a GPU-optimized package version. Oftentimes, investing in a dedicated AI-optimized edge device makes a lot of sense.

■ **Note**   Make sure you are using a GPU-accelerated version of these tools or have a dedicated processing unit for deep learning (more on setting up the environment; see Chapter 3, "Data Scientist's Toolbox").

Surfing: Practical Keypoints Analysis

Occupation? Surfer. Is that a job: surfer? Not really.

—"Surfer Dude," Matthew McConaughey

As a practical introduction to the kind of analysis we can do with keypoints, let's look at surfing. Consider it a surfer's dude guide to machine learning 😊 if you watched the movie with Matthew McConaughey. The sport of surfing is making it to the Olympics for the first time a century after Duke Kahanamoku advocated it to be including in Olympics in 1920. I started surfing for the first time in Hawaii, and after coming back to California, I consistently worked on my skills. I made it all the way from a huge board that looks like an aircraft carrier to a shortboard and I'm learning every day. As a practical example of the sport data science, let's see how we can apply information we can extract from points detected with deep vision methods.

First Look

It's all balance, right? And co-ordination. How hard can it be?

—"Point Break"

Surfing is a sequence of actions that consists of paddling, popping up on your board, riding the wave, turning, and finishing your ride. These actions can be split further into multiple phases and various types of turns: bottom turns, cutbacks, floaters, and more. Having actions split into phases and understanding indicators of each phase helps, as we are going to train our machine learning models to recognize these activities.

PROJECT 7-1: FIRST LOOK AT KEYPOINTS ANALYSIS

Paddling

Just paddling out in a big surf is total commitment.

—"Point Break"

Figure 7-3. Paddling view from a front-view camera and deep vision keypoint detection

Paddling is what surfers do most of the time: they paddle through the waves and as they accelerate to catch up the speed of the wave (Figure 7-3). Have you noticed how lean and shaped surfers' bodies are? This is all because of paddling, one of the most physically demanding activities in the sport.

Figure 7-4. Side view of a surfer paddling

It is a good surfing practice to arch your back to raise on top of the board (Figure 7-4). This helps minimizing resistance and maximizing propulsion, and we may want to measure angles between the joints to see if the surfer needs to arch his/her back more. If we can mathematically estimate angles between body joints, we can use these angles to take a guess about this activity, and to make performance suggestions for the athlete.

How do we practically measure these angles? The keypoints are represented as COCO dataset as points with x and y coordinates and a set of connected body parts, for example, shoulder to elbow. Given three points p0, p1, and p2, for example, representing shoulder, elbow, and wrist, the elbow angle can be calculated as:

```
def angle_calc(p0, p1, p2 ):
    try:
        a = (p1[0]-p0[0])**2 + (p1[1]-p0[1])**2
        b = (p1[0]-p2[0])**2 + (p1[1]-p2[1])**2
        c = (p2[0]-p0[0])**2 + (p2[1]-p0[1])**2
        angle = math.acos( (a+b-c) / math.sqrt(4*a*b) ) * 180/math.pi
    except:
        return 0
    return int(angle)
```

We can use this calculation to estimate angles between three points on the dataset; read on and you'll see how this can be used to analyze more specific poses.

Pop-up

Both feet have to land on the board at the same time.

That's it. That's it. You're surfing.

—"Point Break"

The pop-up (Figure 7-5) is a complex move that involves several actions that happen so quickly with experienced surfers that we may not be able to catch it with our eyes. But with a camera recording at 30 frames per second or more, we can detect that the surfer is getting ready for a pop-up and split it into phases. The pop-up is so essential; practicing it on the ground makes you a better surfer and you are out there in the waves. There're several phases of the pop-up we use in our pose estimation: readiness, transition, and landing.

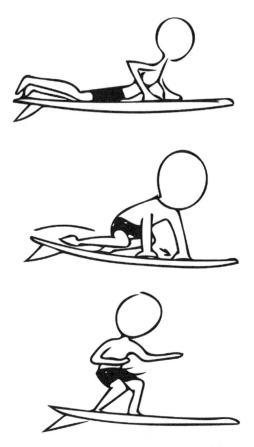

Figure 7-5. Pop-up sequence in surfing

During the first phase, the surfer's back is arched even more than during paddling, head and chin is up, and hands are placed on the rails of the surfboard. The surfer has now caught up with the speed of the wave and is anticipating the moment of pop-up. The second phase, also called the transition, is a quick and efficient movement performed by experienced surfers with grace and fluidity (Figure 7-6). During transition phase, the surfer quickly moves to a classic surfer stance, ready for action.

Figure 7-6. Transition movement in surfing

In the final phase of the pop-up (Figure 7-7), the body is stabilized on the surfboard, and the athlete is ready to rip! This is when the fun starts.

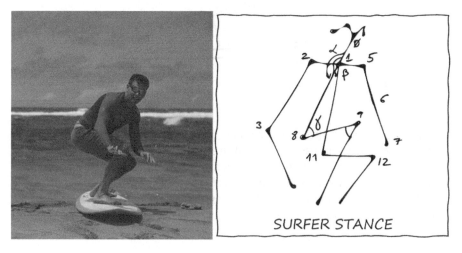

Figure 7-7. Keypoints of the surfer's body during the final pop-up phase

Riding the Surfboard

- What's he searching for? - The ride! The ultimate ride.

—Point Break

Riding the surfboard may involve turning, jumps, and many other tricks. Turns are one of the key elements of surfing. To detect turns using a keypoints estimation output from models, notice key joint angles for athletes. A properly executed turn, for example, with world-class athletes keeps the body compressed, trailing hand nearly touching the water and used as a pivot point, eyes stay on the target, and pressure is on the back foot. Using the approach we used earlier in this chapter, with the set of keypoints, we can detect turns and other tricks.

Surfing is also a lot about meditation, a quiet time with the ocean and nature. And many of us enjoy that quiet time, while sitting on the surfboards, searching for the next wave. The method of body pose estimation with points detected with machine learning models can be applied to many actions and activities.

Beginning a Pose Estimation Project

While I was working on computer vision problems in sports and fitness, I was introduced to the team working with the Gabriel Medina Institute and Rip Curl on surfing research. California summer was getting closer, and being an aspiring surfer myself, I couldn't think of a better application for computer vision. I usually go surfing to Pacifica, my favorite beginner/intermediate break about half an hour away from my house. There, I started playing with a prototype of a smart surfing fin sensor and a custom-built computer camera with onboard computer vision. First things first, in surfing we begin by learning how to stand on the surfboard.

PROJECT 7-2: IS YOUR SURFING STANCE GOOFY OR REGULAR? USING COMPUTER VISION AND 2D POSE ESTIMATION TO DETECT SURFER STANCE

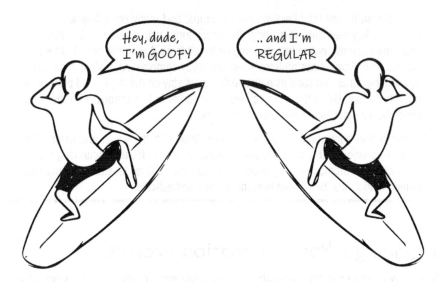

Figure 7-8. A goofy surfer (left) vs. a regular surfer stance (right)

All board sports have a notion of stance (Figure 7-8): be it surfboarding, snowboarding, or skateboarding. The laterality of your brain impacts how footedness is chosen. Naturally, we stay with our left or right foot forward. It's very easy to test without the board by simply noticing what foot we lift first while stepping on the stair (Figure 7-9).

Figure 7-9. How to easily tell a goofy stance from regular

"Goofy" or "regular" means that we step with our right or left leg in front of the board. Wouldn't it be cool if our machine learning could help us recognizing the stance?

Note For how-to on working with code examples and tools, used in this book, please see Chapter 3, "Data Scientist's Toolbox."

To get started, we will use a single image of a surfer in the project folder and 2D pose estimation, to detect athlete's stance; recall the analysis we've done earlier in this chapter. Later we could use a continuous video stream from a webcam, to detect stance continuously in real time.

To estimate the stance, I first infer keypoints and store the inferred points in the pose variable. Next, I pass the pose to the detect_stance method and print the result; you can see that the surfer's stance is identified as goofy (Figure 7-10).

```
print('inference started...')
t = time.time()
pose = e.inference(image, resize_to_default=(w > 0 and h > 0),
upsample_size=4.0)
print("STANCE "+ detect_stance(False, pose))

inference started...
STANCE goofy
```

Figure 7-10. Stance estimation for a surfer

How it works? This example uses TensorFlow for pose estimation. The method I use to determine the stance is by estimating the angle between the head and hips. A greater angle between the left hip and the head means that the athlete is standing in goofy stance, when the camera is not using mirrored mode. Otherwise, I assume that the stance is regular. It's important to notice that in selfies mode, most cameras mirror the image, so to determine stance correctly, we need to pass an argument to the method to tell it whether camera or image is mirrored.

First in detect_stance function, we calculate angles between keypoints and then check stance condition in surfing_check_goofy_stance function.

```
def surfing_check_goofy_stance(is_mirror, a, b):
    '''
        is_mirror shows if the camera is mirrored
        a neck L angle
        b neck R angle
    '''
    if not is_mirror:
        if a > b:
            return BOARD_STANCE_GOOFY
        return BOARD_STANCE_REGULAR
    else:
        if a > b:
            return BOARD_STANCE_REGULAR
        return BOARD_STANCE_GOOFY

def detect_stance(is_mirror, pose):

    l = angle_calc(find_point(pose,0), find_point(pose,1),
    find_point(pose,11))

    r = angle_calc(find_point(pose,0), find_point(pose,1),
    find_point(pose,8))

    return surfing_check_goofy_stance(is_mirror, l, r)
```

To run stance estimation using live webcam real time, check pose_estimation_webcam.py. That script will use your webcam, instead of the image to detect stance; similarly we can use a video file as an input.

Tracking Points Over Time

Fall in love with some activity, and do it!

—Richard Feynman

In the stance detection example, we looked at a very simple problem, by focusing on an individual image or frames. This quick and easy approach results in points detection for a static pose, but since most of the sports are recorded on video, it'd be interesting to see if we can improve precision and recognize entire activities as sequences of events, rather than looking at static poses. For in-depth analysis of video action recognition and deep learning models, check Chapter 9, "Video Action Recognition."

Recognizing activity using rules is useful to understand joint mechanics, but we can also use various algorithms for activity recognition.

Finding Similarities

Coaches often look at athletes performing activities run after run, comparing them with best performing athletes and athlete's own results, then the feedback provided to the athlete that summarizes areas of improvements. Now we have a set of joints, representing athlete's body; it'd be interesting to see how we can compare a similar set of joints over time. Methods like dynamic time warping (DTW) help finding similarities in temporal sequences. For a coach, this is the Holy Grail of sport science! Applying methods that take into account time to sets of activities can help us train models to detect activities as they happen dynamically and, from the coach prospective, detect issues with the movements! For a practical sports scientist looking to analyze athlete movements, this sounds like a great solution. Like all methods these algorithms need to be applied constrained and using sets of data that imply similarity.

PROJECT 7-3: FINDING SIMILARITIES IN POINTS TRAINSITION OVER TIME

Getting Project Data

From project 7-2, you already know how to get a 2D body pose estimation, using either a static image, video, or a webcam. Using tools, such an OpenPose, it's easy to save points into a JSON file, for each frame and each person detected. Conveniently, OpenPose provides methods for saving points out of the box, by using `--write_json` flag:

```
$openpose.bin --model_pose COCO --video examples\media\video.avi --write_json
examples\json
```

The output is a single JSON file per frame, similar to the following, where pose_ keypoints_2 represents an array of 18 keypoint parts locations, followed by confidence value, for example [x1,y1,c1,x2,y2,c2,...]:

```
{
    "version": 1.3,
    "people": [
        {
            "person_id": [
                -1
            ],
            "pose_keypoints_2d": [
                413.834,
                102.427,
                0.896774,
                ...
```

Since we requested a COCO pose model, we'll get 18 points, corresponding to the model we discussed earlier in this chapter. This is generally the data you should be getting from experiments, simulating various activities. The next step is loading this data for comparison.

Basic Idea

Consider two sets of data *x* and *y*, representing a joint movement over time. The idea is to find similarities in movements of a human body joint over time. We will run dynamic time warping method on both sets of points *x* and *y*. This should give us an idea how similar are the activities, or to find discrepancies in technique or execution.

```
data = load_features(['x.npy','y.npy'])
x = data[0]
y = data[1]
plt.plot(x, 'g', label="result")
plt.plot(y, 'r', label="result")
plt.legend();
```

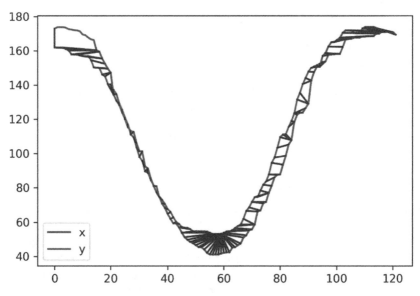

Figure 7-11. Finding similarities with dynamic time warping

It looks (see Figure 7-11) like there's similarity in this data, perhaps both these datasets represent the same activity. One approach to find similarities is finding Euclidean distances, by simply looking at points that can be matched together; less distance means more similarity:

```
distances = np.zeros((len(x), len(y)))
distances.shape
for i in range(len(x)):
    for j in range(len(y)):
        distances[i,j] = (x[i]-y[j])**2
distances
```

We can use a Python module called FastDTW to estimate the distance between points and plot that dependency. The notebooks included with the source code of this chapter include additional methods for visualization of the cost matrix and the algorithm:

```
from scipy.spatial.distance import euclidean
from fastdtw import fastdtw
distance, path = fastdtw(x, y, dist=euclidean)
print(distance)
```

```
path_x = [point[0] for point in path]
path_y = [point[1] for point in path]
plt.plot(path_x, path_y)
plt.show()
```

Finding similarities in temporal data is a common task in data science, and as demonstrated in this example, it can be applied to the set of human joints detected with deep vision methods.

Detecting a Skill Level

Detecting a skill level in sports is one of the most important problems for a coach: from the moment we see an athlete, neural networks in our brain attempt to classify movements of that athlete, analyzing how good of an athlete that is. Perhaps, this natural function of our brain is related to our basic survival instincts: this classification is trained to be superfast, in an attempt to protect us from unfriendly or dangerous acts. Since ancient times, we had to quickly estimate the level of danger; with the evolution of human society, we instinctively look for individuals who we can learn from, by classifying or comparing our own skill to theirs.

Most modern sports developed comprehensive systems that define skill levels. For example, the Professional Ski and Snowboard Instructors of America have defined ten skill levels that help classify skiers and recommend course of training. Generally, most sports classify athletes' skill as beginner, intermediate, and advanced (experts). In our next project, we will use pose estimation to tell an advanced athlete from a beginner.

PROJECT 7-4: USING POSE ESTIMATION TO DETECT A SKILL LEVEL

For this project we'll dive into skiing, a highly dynamic sport. Even if you are not a skier, by using data from pose estimation, you'll be able to use methods described earlier to tell an advanced level athlete from a beginner (Figure 7-12).

Figure 7-12. Athlete with advanced skill level: body pose points

Let's look at a beginner-level skier, in this case a kid skiing wedge type of turn in a typical beginner stance (Figure 7-13). Apart from balance and center of mass being impacted by body proportions (for kids, their head is bigger relative to the body), a professional instructor can immediately identify a number of elements typical for beginners. Let's see if we can use machine learning methods to help us identify those and predict skill levels of athletes. Looking at the body position, notice that the center of mass of the skier is far back, instead of being on top of the center of the skis. Angle between torso and hips tends to be high; the kid is "sitting back," which is very typical for kids skiing but is also characteristic for beginner adult skiers.

SKILL DETECTION

BEGINNER ATHLETE   A-FRAME POSE

Figure 7-13. Keypoints and a typical pose of a beginner skier

Note also the neck angle: beginners tend to look at the ski tips, instead of looking straight, about three turns ahead; as advanced skiers do, the neck is also extended forward to compensate for the backward body position. Many beginners tend to use upper body to turn, instead of using ankles and knees. On the video sequence (Figure 7-14), it's clear that what's happening here is hips, ankles, and knees stay static, locked in the so-called A-frame pose, while the upper body swings, providing a momentum for turn.

Figure 7-14. Beginner skier visualization sequence with a typical A-pose

We will use an advanced skier dataset and a dataset from the beginner skier to compare and predict the skill level based on differences in body joints relations. First, we get neck and torso vectors from body parts:

```
# Looking down at skis? check neck and torso
    neck = np.array([(joint[5].x - joint[4].x, joint[5].y - joint[4].y)
    for joint in joints])
    torso = np.array([(joint[4].x - joint[3].x, joint[4].y - joint[3].y)
    for joint in joints])
```

Let's normalize these vectors and calculate joint angles:

```
# Normalize vectors
    neck = neck / np.expand_dims(np.linalg.norm(neck, axis=1), axis=1)
    torso = torso / np.expand_dims(np.linalg.norm(torso, axis=1), axis=1)

    # Calculate angles
    angles = np.degrees(np.arccos(np.clip(np.sum(np.multiply(torso, neck),
    axis=1), -1.0, 1.0)))
    print("Max angle: ", np.max(angles))
```

The resulting plot shows that the beginner skier has upper body rotating and swinging a lot more than an advanced athlete (Figure 7-15).

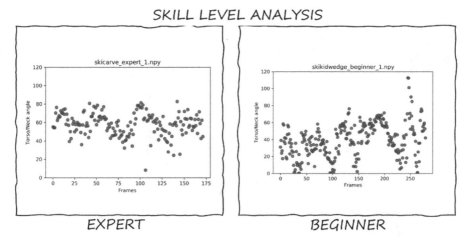

Figure 7-15. Expert vs. beginner skill-level analysis: torso/neck during the movement

This result tells us what a ski instructor could explain in a few seconds by observing both skiers: quiet upper body of a more advanced athlete, and turns initiated from ankles and knees, vs. swinging the body in case of a beginner. By detecting a set of points on the human body with deep vision models or by using sensors, we can analyze performance metrics specific to each sport.

Multi-person Pose Estimations

Many sports involve a multiplayer action, so it is important to detect groups of people. Fortunately, most methods we discussed earlier for 2D and 3D body pose estimations apply for detection in the group. One of the most important aspects of group detection for sports scenarios is performance (earlier we discussed top-down and bottom-up approaches to pose estimation).

Figure 7-16. Multi-person pose estimation

Fortunately for sport data scientists, 2D pose estimation works really well in a multi-person scenario (Figure 7-16). In the following project, we will use a simple people count, detected from pose estimation.

PROJECT 7-5: MULTI-PERSON POSE ESTIMATION AND PEOPLE COUNTING

To count people detected from the inference, simply use len(people) to get the size of the array from the inference:

```
people = e.inference(image, resize_to_default=(w > 0 and h > 0),
upsample_size=4.0)

count_people = len(people)

print("NUMBER OF PEOPLE ", count_people)
```

Once we have a group of people detected, we could apply prediction methods to detect information about specific subjects in the group.

Dealing with Loss and Occlusion

Occasionally, models may exclude points that are completely hidden from the observer (keypoint probability drops to 0.0 in our data). As a practical data scientist, you can deal with the problem in many ways. Depending on the activity, a data scientist may have an option of installing additional cameras with additional view angles to reduce the number of occluded parts. This method is called multiview video triangulation. Depth sensors, such as LIDARs, may provide additional information, as in the case mentioned earlier in this book with AI judging system in gymnastics. If the activity you're studying allows the use of additional sensors (such as IMUs) and the activity allows placing sensors on athlete's body, inertial movement sensors don't suffer from optical occlusion and are discussed in other parts of this book, dedicated to sensors and hybrid methods.

One simple method, based on the datasets we are dealing with, is including only the frames with joints present across all frames. This approach is used in get_joints method defined in utils.py in projects for this chapter:

```
joints = [joint for joint in poses if all(part.exists for part in joint)]
```

Despite occlusion, pose estimation remains a very robust, practical, and near real-time method for movement analysis in sports.

Summary

In this chapter we looked at 2D body pose estimation applications for sports. Pose estimation methods and models have reached maturity and precision in many areas, while they still don't generalize well just yet, as research shows, they can be applied to specific activities in sports. We demonstrated some examples and applications in practical projects included in this chapter, with highly dynamic sports, including surfing and skiing. Pose estimation can be applied in sports movement analysis if you are willing to accept occasional instances of occlusion and exclude frames with loss of data points. In the next chapter, we extend techniques we used in 2D body pose estimation to three-dimensional models. For additional materials and a video course, supplementing this book, check my Web site ActiveFitness.AI http://activefitness.ai.

3D Pose Estimation

Who would believe that so small a space could contain the image of all the universe? O mighty process!

—Leonardo da Vinci, Camera Obscura

Figure 8-1. Surfer pose estimation with dense surface

© Kevin Ashley 2020
K. Ashley, *Applied Machine Learning for Health and Fitness*,
https://doi.org/10.1007/978-1-4842-5772-2_8

Overview

In the previous chapter, you focused on estimating keypoints of human body poses, mostly based on a flat images. In other words, our goal was reconstructing joints of the human body, without paying attention to analyzing how these joints are positioned in space. The task of estimating position of objects and human body in three-dimensional space is more challenging, but the result is stunning (Figure 8-1)! Imagine that you are dealing with a simple RGB camera: the camera works by projecting light it captures in the scene of the real three-dimensional world into a sensor, generating a 2D view of the world. For sport scientists, a camera is the most obvious source of information, but it's not necessarily the most accurate.

Figure 8-2. Kiteboarder pose estimation using DensePose

First, given camera line of sight, some parts of the scene may be hidden from the observer. Also, the camera is limited to a certain rate of frames per second it can capture; devices such as sensors significantly exceed that rate. Having extra sensors, such as depth (e.g., LIDAR, introduced in Chapter 5, "Sensors") and infrared, as well as motion sensors, such as inertial measurement units (IMUs), is often a luxury in the field. Getting these sensors installed in a live sport action environment (Figure 8-2) or on the athlete's body is often impractical, so machine learning often relies on using methods such as transfer learning or ground truth data for training. A data scientist can train a model using a ground truth dataset, such as Human3.6m, including data collected from sensors, and then let the model infer position from the video.

Despite the massive use of video, hybrid methods combining positioning, acceleration or depth sensors, and video data are still widely used where conditions permit. A notable example of such approach is used by the International Federation of Gymnastics and Fujitsu in Gymnastics competition judging AI, which uses a combination of a LIDAR and a camera to provide a near real-time 3D body pose reconstruction. Methods of pose reconstruction in 3D vary, depending on the inputs. In this chapter we'll investigate methods of reconstructing 3D based on various sources of information: single cameras, multiview cameras, and triangulation and sensors to estimate pose more accurately in 3D space.

In this chapter, we'll start with fundamental principles of camera optics that will help us with 2D-3D reconstruction, to more advanced questions of how our eyes can reconstruct depth in the scene and then using machine learning.

Cameras and 3D World

The creation of a single world comes from a huge number of fragments and chaos.

—Hayao Miyazaki

A pinhole camera, as a device that maps from 3D world into a 2D image plane, has been known for thousands of years and is mentioned by many ancient writings, including Aristotle, Chinese, and Indian. A mathematical model of an ideal pinhole camera, that is, the camera without lenses and aperture of a single point, is often used in modeling optics. Mathematically, a pinhole camera matrix C makes the mapping by projecting world space points X into the image plane Y:

$$Y = CX$$

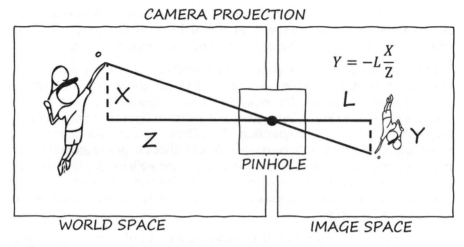

Figure 8-3. Simplified pinhole camera projection model

In Figure 8-3, you can see two similar triangles, with hypotenuses representing projection lines from the hand of the tennis player in the world space to the similar point on the projected image inside the camera on the sensor. Since both triangles are similar:

$$-\frac{Y}{L} = \frac{X}{Z} \Rightarrow Y = -L\frac{X}{Z}$$

This transformation can be written in terms of *projective coordinates* from the world space to image plane. In 3D space, we can write down such a transformation in a matrix form (note that the following formula applies to an ideal pinhole camera, without a lens):

$$\begin{bmatrix} y_1 \\ y_2 \\ 1 \end{bmatrix} = \begin{bmatrix} f_x & 0 & 0 & 0 \\ 0 & f_y & 0 & 0 \\ 0 & 0 & 1 & 0 \end{bmatrix} \begin{bmatrix} x_1 \\ x_2 \\ x_3 \\ 1 \end{bmatrix}$$

Unlike pinhole cameras, actual cameras typically include a lens, and a lens may introduce some distortion. Most cameras distort images (Figure 8-4)—for example, when you take a picture for your Facebook account, and you end up with something that looks less than ideal because your face may look stretched on the sides. Or, if you take a picture of a document, and it looks unreadable.

Some applications like Microsoft Office Lens can undistort documents, and OpenCV includes undistort method. How's that done? Optical distortion occurs because the camera takes the light coming from a 3D scene of the world and forms a 2D image on the sensor, after that light comes through the lens. On some images, straight lines appear curved, and some objects may appear tilted, closer, or farther away. It may not be a big deal for your Facebook or Instagram account, but it's a bigger problem for self-driving cars that need to estimate the curvature of the road. Tangential distortion occurs when the lens is angled relative to the sensor. Radial distortion occurs because the light bends more toward the edges of the lens than in the center.

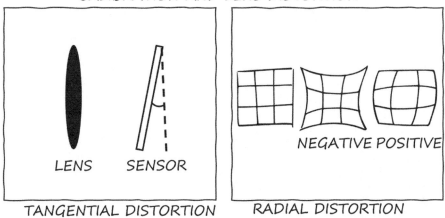

Figure 8-4. Camera lens distortion effects

Camera Matrix

To solve the distortion problem and find a camera-specific calibration matrix and distortion coefficients, we can use computer vision libraries, such as OpenCV. Calibration matrix can be later saved to a file for later reuse as a NumPy array; this may be useful to persist camera calibration information. To calibrate, OpenCV offers calibrateCamera method that takes points from 3D object space and points from 2D image space and outputs a camera matrix that we can later reuse for calibration. In this section we are going to show how to do this in Python with OpenCV.

PROJECT 8-1: CAMERA MATRIX AND 3D RECONSTRUCTION

Typically, to calibrate the camera, a grid, like a chessboard, is used. You can take multiple pictures of a chessboard with your camera at different angles or use a set of images such as the ones in data/camera/chessboard3 folder (usually 10–20 pictures at different angles are sufficient). I included a helper method in `utils.pose3d` that includes methods for camera calibration, based on OpenCV:

```python
import cv2
import pickle
import glob
import matplotlib.pyplot as plt
from utils.pose3d import pose3d
%matplotlib inline

images = glob.glob('data/camera/chessboard3/*.jpg')
points_3d, points_2d = pose3d.find_corner_points(images)
img = cv2.imread(images[0])
# Calibrate
_, M, D, R, T = cv2.calibrateCamera(points_3d, points_2d,
(img.shape[1], img.shape[0]), None, None)
dist_pickle = {}
dist_pickle["mtx"] = M
dist_pickle["dist"] = D
# Save calibration
pickle.dump(dist_pickle, open( "data/camera/calibration.pickle", "wb" ))
```

In this example, I use `calibrateCamera` method from OpenCV, which returns a camera matrix M, a vector of distortion coefficients D, the vector of rotation R, and the vector of translation T. We can save camera matrix and distortion vector in the pickle file and reuse them later, because they are specific to the camera. Next, we can plot axis of a reconstructed 3D space based on the camera matrix (Figure 8-5):

```python
img = pose3d.find_3d_cube(images[0],M,D)
plt.imshow(img)
plt.show()
```

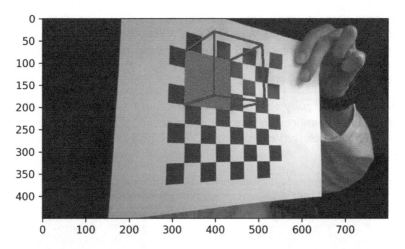

Figure 8-5. Reconstructed 3D space and axis

In this example you walked through camera calibration process and reconstructing a 3D view based on a special pattern, like a checkerboard, after you calculated camera matrix and distortion.

Using a Single Camera

A single-view camera can be used to reconstruct a 3D human body pose using a convolution neural network. Most models use a set of 2D keypoints detected using models we discussed in the previous chapter and then apply a regression model to map 2D keypoints back to 3D space (Figure 8-6).

SINGLE CAMERA 3D RECONSTRUCTION

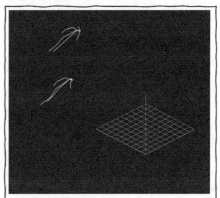

2D KEYPOINTS   3D RECONSTRUCTION

Figure 8-6. Reconstructing 3D pose from a video of a single 2D camera

For a sport data scientist, this capability to use 2D video data to analyze human body in 3D space is very valuable. While keypoints can be used for pose representation, a more accurate estimation of fully blended 3D body shapes can also be done with models like SMPL (Skinned Multi-Person Linear model), as illustrated in Figure 8-7.

Models for human body reconstruction in full 3D including joints or shapes have achieved an impressive performance recently, with the mention of a notable SMPLify method estimating 3D body pose and shape from a single image!

SINGLE CAMERA 3D HUMAN INFERENCE WITH SHAPE

VIDEO   3D SHAPE RECONSTRUCTION

Figure 8-7. SMPL 3D body shape inferred from a single camera video

In the next section, we'll go over another method that applies 3D triangulation from multiple views, similarly to what our eyes are using to reconstruct depth in the scene.

Multiview Depth Reconstruction

There's a reason why humans and many other biological forms on our planet have a pair of eyes. In modern science many insights come from our understanding of how we perceive the world, and multiview vision is one of the most interesting insights about vision data. A single image carries a lot of information, but it's missing a key data point that our brain needs to accurately reconstruct 3D images: depth! Without depth, any image is ambiguous, that is, it makes it hard for our brain to perceive distance to objects. Knowing depth makes it easier to estimate distance in the visual scene, and that is a critical component to our understanding of the 3D environment.

DEPTH TRIANGULATION

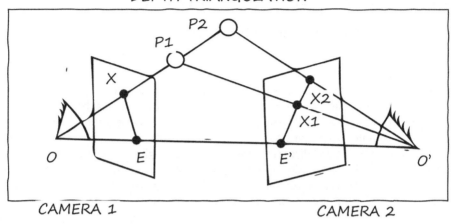

Figure 8-8. Multiview triangulation

In fact, what our brain is doing with two eyes is called triangulation. It's very easy to explain: with one eye, using pinhole camera model, we only have one image plane; every point from the 3D world along the straight line projects to a single point on the image plane! With the second eye, looking at the line from a slightly different angle, we can see multiple points, so our brain can triangulate 3D points! In Figure 8-8 (multiview triangulation), the world is observed from two points of view, O and O'. If you only look at points P1 and P2 with one eye O, you will not see the difference between these points. This is because, for one eye, these points appear as one point X. With the second eye O', we can create a plane XOO' called *epipolar plane* and project points on OX to the second image plane. Line XE corresponding to point X is called *epiline*. How do we find the matching point on the second image? This task is simplified because of what's called an *epipolar constraint*: to find a matching point, we don't need to search all points on the second image plane; we can only search points along the epiline. But because the second eye O' is at an angle, instead of one point we can see multiple points (in our case X2 and X1) on the second *epiline*! This method allows optically reconstructing one of the most important aspects of 3D imaging, depth.

In the next sections, I'll discuss reconstructing a 3D object with sensors, a method often used in motion caption processing. The multiview 3D triangulation we discussed earlier is used in practical frameworks, and requires a setup with multiple cameras.

3D Reconstruction with Sensors

I thought 2-D and 3-D could coexist happily.

—Hayao Miyazaki

In addition to vision, sensors such as inertial movement units (IMUs) can be used to provide data for 3D pose reconstruction. Well-calibrated sensor units, such as Xsens, can provide a high-quality motion capture that can be used as a ground truth data for training models, or to analyze and visualize human body pose in 3D directly, often with much higher precision than vision data. We already discussed why using inertial movement sensors may be beneficial for 3D data capture, but it's worth reiterating the main reasons in the context of using sensor data in 3D reconstruction:

- Sensor data, such as IMU, unlike vision-only information often provides full 3D transformation, including acceleration, rotation, and displacement.

- Sensor data in the IMU typically comes at a very high rate (often 100 or 200 Hz), while visual data is typically capped at 30–60 fps.

- Range-measuring sensors, such as LIDARs, can also be used to provide depth information.

Motion Capture

If you can dream it, you can do it.

—Walt Disney

Motion capture (mocap) has emerged from studio animation and simulation domain, to become a valuable tool for collecting high-quality datasets for machine learning. Traditionally, mocap is used to create high-quality studio animations and animated sequences for interactive applications, movies, games, and simulations and in sports for high-precision analysis of athletic movements. Many sports organizations, including the US Ski Team, NBA, NFL, WTA, and Professional Ski and Snowboard Instructors of America, apply motion capture and video analytics (Figure 8-9). Mocap data is often used in machine learning datasets as a high-quality ground truth data, which includes measurements from sensors, supplementing visual information.

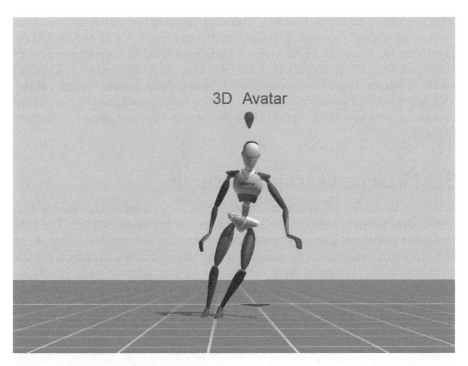

Figure 8-9. Capturing skiing action with Xsens full-body suit

3D Datasets

Let's spend some time discussing datasets available for 3D pose estimation. These datasets may use different motion capture methods, size, and categories of actions. Checking assumptions made during data collection of these datasets is important if you are planning to use models trained with this data or train your own models. Some datasets include ground truth data, the actual measurements from motion capture systems and sensors. Some of the most widely used datasets used to train 3D models are:

- Human3.6m contains 3.6 million 3D human images taken from several cameras and high-speed mocap sensors. The dataset is organized into several categories corresponding to various actions.

- HumanEva dataset contains video sequences synchronized with body points taken from a motion capture system.

There're other datasets available, including those based on synthetic data, such as games, for example, JTA Dataset (Grand Theft Auto) and SURREAL (Synthetic hUmans foR REAL tasks). DensePose-COCO dataset is a part of Facebook DensePose project that has 50K humans annotated with 5 million labels corresponding to body parts. Leeds Sports Pose dataset contains about 2K images with 14 joint positions. For training models including shapes, some recent datasets use SMPL, like AMASS (Archive of Motion Capture as Surface Shapes).

3D Machine Learning Methods

A human 3D body pose reconstruction (Figure 8-10) is more challenging than a two-dimensional approach we've done in the previous chapter. The first pages of this chapter gave you some fundamentals of 3D reconstruction, from the simplest pinhole camera, calibrating and using libraries such as OpenCV to reconstruct 3D space, to basics of depth triangulation with multiview cameras. As sport data scientists, our primary interest is of course in analyzing and reconstructing a human body. This is a much more complicated task, and it turns out that machine learning can help (Figure 8-11).

Figure 8-10. NBA action shows capturing groups of people

Figure 8-11. Dense pose estimation of a flying kiteboarder

Some models use two-dimensional detections as input for the 3D pose estimations. What came as a rather pleasant surprise for data scientists in the recent years of research is the low error rates of lifting ground truth body joint locations to predict 3D positions! It turns out that once two-dimensional joint locations have been detected (like you did in the previous chapters' projects!), it's possible to design a network that uses joint angles and bone lengths for human body inference in 3D space. This approach typically involves a two-step process, where each step is separate: in the first step, the two-dimensional keypoints are detected; in the second step, the model performs the inference in 3D space. In more recent research, it was shown that a 2D-3D transfer learning can be more efficient, when the layers detecting two-dimensional keypoints directly connect with the depth-regression layer in the same network.

For many sports, it is important to detect groups of people. Fortunately, most models for 2D and 3D body pose estimations apply for detection in the group. One of the most important aspects of group detection for sport scenarios is performance.

Sparse and Dense Reconstruction

Figure 8-12. Joints (left) and surface (right) pose representation

Sparse reconstruction means providing a limited set of key points or joints; *dense* reconstruction applies to methods that detect surface areas (see Figure 8-12). When using a model to represent a human body, you have a choice of models based on relatively sparse joints representation (usually 17 or 25 joints) and those using mesh or surface. In the previous chapter, we used a keypoint detection method, applying a built-in PyTorch R-CNN ResNet FPN model trained on COCO dataset classes to detect keypoints of a surfer (Figure 8-13).

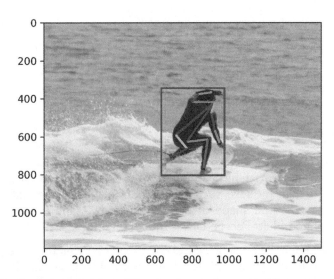

Figure 8-13. Sparse keypoints from 2D estimation can be used in 3D reconstruction

```python
from torchvision import models
from PIL import Image
import matplotlib.pyplot as plt
import torch
import torchvision.transforms as T
%matplotlib inline

img = Image.open('./media/surfer.jpg')
plt.imshow(img); plt.show()

def normalize(img):
    normalize_t = T.Compose([T.Resize(256),
                    T.CenterCrop(224),
                    T.ToTensor(),
                    T.Normalize(mean = [0.485, 0.456, 0.406],
                                std = [0.229, 0.224, 0.225])])
    return normalize_t(img).unsqueeze(0)

img_tensor = normalize(img)

kprcnn = models.detection.keypointrcnn_resnet50_fpn(pretrained=True)
kprcnn.eval()
detection = kprcnn(img_tensor)[0]keypoint_detections =
get_keypoint_detections(img)
json_keypoints = to_json(keypoint_detections)
print(json_keypoints)
```

Which results in these key points detected:

```
'[{"nose": [924, 420, 1], "left_eye": [923, 408, 1], "right_eye": [920, 405,
1], "left_ear": [884, 404, 1], "right_ear": [895, 405, 1], "left_shoulder":
[803, 415, 1], "right_shoulder": [910, 478, 1], "left_elbow": [789, 408, 1],
"right_elbow": [924, 562, 1], "left_wrist": [954, 636, 1], "right_wrist":
[951, 636, 1], "left_hip": [723, 580, 1], "right_hip": [751, 610, 1], "left_
knee": [834, 570, 1], "right_knee": [887, 613, 1], "left_ankle": [867, 753,
1], "right_ankle": [865, 759, 1]}]'
```

This approach of detecting keypoints in 2D is used as a base for 3D pose regression in many machine learning models. Another type of models also often called dense models, such as Facebook's DensePose, takes another approach and uses the shape of the human body. Many dense models are based on Skinned Multi-Person Linear model (SMPL).

The following example is using DensePose to infer an athlete dense pose using a prebuilt model available with the framework. The inference is done with infer_simple.py tool that comes as part of the toolkit. This call takes an image of an athlete and generates two images that contain I,U,V channels with resulting segments detected by the model, which makes it easy to overlay the inferred shape against the actual source image (Figure 8-14).

Figure 8-14. Segments captured with DensePose

```
python2 tools/infer_simple.py --cfg configs/DensePose_ResNet101_FPN_s1x-e2e.
yaml --output-dir DensePoseData/infer_out/ --image-ext jpg --wts
https://dl.fbaipublicfiles.com/densepose/DensePose_ResNet101_FPN_s1x-e2e.pkl
DensePoseData/demo_data/surf4.jpg
```

The arguments for the script include input image, model weights, a configuration for the model, and output directory. If you want to visualize these images, this snippet helps combining resulting channels:

```
fig = plt.figure(figsize=[30,30])
plt.imshow( np.hstack((IUV[:,:,0]/24. ,IUV[:,:,1]/256. ,IUV[:,:,2]/256.)) )
plt.axis('off') ;
plt.show()
```

This resulting segment information can also be used to draw a contour of the body:

```
fig = plt.figure(figsize=[30,30])
plt.imshow( im[:,:,::-1] )
plt.contour( INDS, linewidths = 4 )
plt.show()
```

As part of TensorFlow Graphics, tensorflow library recently introduced a new method that uses mesh semantic segmentation of a 3D mesh. In describing the method used for this segmentation, TensorFlow refers to a method called Feature-Steered Graph Convolutions for 3D Shape Analysis. TensorFlow also conveniently provides a tool inside TensorBoard that visualizes the 3D mesh.

Summary

Human pose estimation methods using machine learning have significantly advanced in the last few years, triggering a revolution in many industries, including gaming, fashion, sports, and robotics. We now have better ways to collect ground truth data with sensors and synchronized video, as part of the motion capture (mocap). And while motion capture for sports has elements of art and science at the same time and understanding mocap tools and limitations of each method of motion capture still comes with practice and expertise, a data scientist today has datasets available for research. In this chapter we discussed basic principles of camera optics and fundamentals of 3D reconstruction with a monocular and multiview camera. We also discussed datasets and some of the most advanced methods for 3D pose reconstruction available today and discussed sparse and dense reconstruction. I hope from reading this chapter you had a good overview of some of the basics, as well as more advanced machine learning techniques. For additional materials and a video course, supplementing this book, check my Web site ActiveFitness.AI http://activefitness.ai.

Video Action Recognition

If a picture is worth a thousand words, then a video is worth a thousand times more.

Figure 9-1. Sequence of video frames shows correlation between frames

© Kevin Ashley 2020
K. Ashley, *Applied Machine Learning for Health and Fitness*,
https://doi.org/10.1007/978-1-4842-5772-2_9

Background

Our brain is a superfast action recognition system that's hard to match. In terms of deep learning, our brain routinely does many things to recognize actions, and it works *fast*! It may be years of evolution, the need to identify incoming danger or provide food: each of us has a miraculously fast video action recognition engine that just works. Our brain is capable of *normalization* and *transformation* since we recognize actions regardless of viewpoint. Human brain is great at *classification*, telling us what's moving and how, and it can also *predict* what's coming next. It turns out, our knowledge of deep learning for action recognition is getting close, but it's not as good just yet. Especially, it becomes clear when generalizing movements: the brain is far ahead of neural nets in its ability to generalize. Although if data science keeps the same pace of evolution as in the last decade, perhaps AI may get closer to the human brain.

ACTION SEQUENCE

Figure 9-2. Recognizing tennis play from this sequence is easy for human brain

Recognizing actions from videos is key to many industries (Figure 9-1): sports, security, robotics, health care, and many others. For a practical sport data scientist or a coach, action recognition is part of the daily job: a coach's eye and experience is trained to do movement analysis (Figure 9-2).

In biomechanics, we use physical or classical mechanics models to describe movement. This approach worked for many years in sport science, but analytical methods are complicated and as history shows, although movements can be described with classical mechanics, deep learning methods can be more efficient and often demonstrate great precision. To illustrate this point, Kinetics dataset described in this chapter contains 400 activity classes with hundreds of sport activities recognition readily available. You can classify any of those activities with an average-level computer, and you really don't need a PhD in biomechanics.

Kinetics dataset has half a million video clips, covering hundreds of human actions and totaling close to a terabyte of data. That looks like a lot, but for each sport, it covers only a few basic moves. A human ski coach evaluating a skier can narrow it down to dozens of small movements and usually deals with multiple training routines.

Video recognition has been traditionally tough for deep learning because it needs more compute power and storage than most other types of data: that's a lot of power and storage! In the recent years, video recognition methods, models, and datasets made a significant progress to the point that they became practical for a sport scientist. As this chapter shows, these methods are also relatively easy to use with the tools, frameworks, and models available today. In this chapter, we focus on practical use and methods for video action recognition. You don't need an advanced hardware, but a GPU-enabled computer is recommended. If you don't have that handy, use an online service like Microsoft Azure or Google Colaboratory that offers free scaleable compute services for data science.

Video Data

Cinematography is a writing with images in movement and with sounds.

—Robert Bresson, *Notes on the Cinematographer*

Video classification has been an expensive task because of the need to deal with the video, and video is heavy. In this chapter you'll learn data structures for video, used across most of datasets and models for video recognition.

A single image or video frame can be represented as a 3D tensor: (width, height, color) and color depth having three channels, RGB. A sequence of frames can be represented as a 4D tensor: (frame, width, height, color). For video classification, you will typically deal with sequences, or batches of frames, and the video is represented as a 4D or 5D tensor: (sample, frame, width, height, color).

For example, to read a video into structures ready for deep learning, frameworks provide convenience methods, such as PyTorch's `torchvision.io.read_video`. In the following code snippet, the video is loaded as a 4D tensor 255 (frames) x 720 (height) x 1280 (width) x 3 (colors). Notice that this method also loads audio, although we are not going to use it for action recognition:

```
import torchvision.io
video_file = 'media/surfing_cutback.mp4'
video, audio, info = torchvision.io.read_video(video_file, pts_unit="sec")
print(video.shape, audio.shape, info)
```

```
Output:
torch.Size([255, 720, 1280, 3]) torch.Size([2, 407552]) {'video_fps':
29.97002997002997, 'audio_fps': 48000}
```

This original video at 720p resolution is large; most models are trained with images and videos that are much smaller in size and were normalized.

Datasets

The relationship between space and time is a mysterious one.

—Carreira, Zimmerman "Quo Vadis? Action Recognition"

Quo Vadis is Latin meaning "where are you going?" Although action recognition is achievable from a still frame, it works best when learning from temporal component as well as spatial information.

From the still frame in Figure 9-3, it's not easy to tell whether the person is swimming or running. Perhaps, this ambiguity in action recognition prompted authors of research article on Kinetics video dataset and the model they developed to choose the title for the research work. Prior to Kinetics, Sports-1M used to be a breakthrough dataset used in many models, and in authors' own words:

> *To obtain sufficient amount of data needed to train our CNN architectures, we collected a new Sports-1M dataset, which consists of 1 million YouTube videos belonging to a taxonomy of 487 classes of sports.*

> —Andrej Karpathy et al., "Large-Scale Video Classification with Convolutional Neural Networks"

QUO VADIS?

Figure 9-3. Quo Vadis? Why is the article on Kinetics dataset and ConvNets named after 1951 epic?

Historically, deep learning for video recognition focused on activities that were easily available. What source of video data can a data scientist use without the need to store terabytes of videos? YouTube comes very handy, as well as any other online video service! You'll notice that many of the action recognition datasets use online video services because those videos are typically indexed, can be readily retrieved, and often have additional metadata that helps classifying entire videos or even segments. In fact, with massive online video repositories, storing billions of movements and deep learning, we are on the verge of revolution in movement recognition!

Some well-known datasets for human video action sequences include:

- **HMDB 51** is a set of 51 action categories, including facial and body movements and human interaction. This dataset contains some sport activities, but is limited to bike, fencing, baseball, and a few others. This is included in PyTorch: `torchvision.datasets.HMDB51`.

- **UCF 101** is used in many action recognition scenarios, including human-object interaction, body motion, playing musical instruments, and sports. This is included in PyTorch: `torchvision.datasets.UCF101`.

- **Kinetics** is a large dataset of URL links to video clips that covers human action classes, including sports, human interaction, and so on. The dataset is available in different sizes: Kinetics 400, 600, and 700 and is included in PyTorch: `torchvision.datasets.Kinetics400`.

Models

While video presents many challenges, such as computational cost and capturing both spatial and temporal action over long periods of time, it also presents unique opportunities in terms of designing data models. Over the last few years, researchers experimented with various approaches to video action recognition modeling. Methods that prove most effective so far are using pretrained networks, fusing various *streams* of data from video, for example, motion stream from optical flow and spatial pretrained context (Figure 9-4).

Figure 9-4. Modern models use fusion of context streams: for example, temporal and spatial for action recognition

Some earlier methods tried experimenting and benchmarking model performance with various context streams, for example, using different context resolutions with a technique authors creatively called a "fovea" stream in one research combined with the main feature learning stream.

> *Fovea – a small depression in the retina of the eye where visual acuity is the highest.*
>
> —Oxford Dictionary

This area of deep learning is still under active research and we may still see state-of-the-art models that outperform existing methods.

Video Classification QuickStart

PROJECT 9-1: QUICK START ACTION RECOGNITION

This project provides a quick start for video classification: the goal is to have a practical sport data scientist quickly started on the human activity recognition. Before we start on this project, let's take a look at the list of human activities we can classify with minimal effort. I'll be using PyTorch here, because it provides video classification datasets and pretrained models out of the box. PyTorch computer vision module, torchvision, contains many models and datasets we can use in sports data science, including classification, semantic segmentation, object detection, person keypoint detection, and video classification. Video classification models and datasets included with PyTorch are what we'll be using for this task to get started quickly.

In PyTorch video classification models are trained with Kinetics 400 dataset. Although not all of these human activities are sports related, in the source code I put together a helper in `utils.kinetics`; conveniently, it provides a list of sport-related activities:

```
from utils.kinetics import kinetics
categories = kinetics.categories()
classes = kinetics.classes()
sports = kinetics.sport_categories()

count = 0
for key in categories.keys():
    if key in sports:
        print(key)
        for label in categories[key]:
            count+=1
            print("\t{}".format(label))
print(f'Sport activities labels: {count}')

Output:
Sport activities labels: 134
```

■ **Note for activity granularity**   The trend with activity recognition is getting even more granular. For example, in golf, Kinetics 400 classifies chipping, driving, and putting and, for swimming, backstroke, breast stroke, butterfly, and so on.

So, from several video classification models available in torchvision pretrained on Kinetics 400 dataset, we should be able to get 130+ sports-related actions classified. To quick start our development, we will jump start action recognition with pretrained models available in PyTorch torchvision. Let's start by importing the required modules:

```
import torch
import torchvision
import torchvision.models as models
```

Since we are dealing with models that involve video tensors, having a CUDA-enabled installation of PyTorch really helps. Processing on a CPU will work, but may take a very long time. To make sure we are using a CUDA device, run this command:

```
device = torch.device("cuda" if torch.cuda.is_available() else "cpu")
print(device)
```

```
Output:
cuda
```

Then, get an appropriate model trained on Kinetics 400. Currently, PyTorch supports three video classification models out of the box: ResNet 3D, ResNet Mixed Convolution, and ResNet (2+1)D. I instantiated ResNet 3D (r3d_18) and commented out two other models. The important thing of course is `pretrained=True` flag that saves us downloading all the videos for Kinetics 400 dataset to train the model!

```
model = models.video.r3d_18(pretrained=True)
#model = models.video.mc3_18(pretrained=True)
#model = models.video.r2plus1d_18(pretrained=True)
model.eval()
```

Thanks to PyTorch magic, the pretrained model gets downloaded automatically, a huge time saver! Training on Kinetics 400 dataset requires a massive number of videos downloaded. So, having a pretrained video classification model in PyTorch is a great starter for a practical sport data scientist.

Practical tip Having a pretrained model for Kinetics in PyTorch torchvision offers a big advantage from the practical standpoint. Downloading datasets and videos to train video classification models takes a lot of space and compute power!

Next, we need to define our normalization for the video. For Kinetics, height and width are normalized to 112 pixels, the mean to [0.43216, 0.394666, 0.37645], and standard deviation to [0.22803, 0.22145, 0.216989], parameters recommended in the pretrained model. In the code snippet that follows, this normalization is implemented in several methods that we will later apply in sequence to convert the video to PyTorch tensor, resize and crop it, and normalize to the mean and standard deviation, so that it matches the model:

```
# Normalization for Kinetics datasets

mean = [0.43216, 0.394666, 0.37645]
std = [0.22803, 0.22145, 0.216989]

# To normalized float tensor
def normalize(video):
    return video.permute(3, 0, 1, 2).to(torch.float32) / 255

# Resize the video
def resize(video, size):
    return torch.nn.functional.interpolate(video, size=size, scale_
    factor=None, mode="bilinear", align_corners=False)

# Crop the video
def crop(video, output_size):
    h, w = video.shape[-2:]
    th, tw = output_size
    i = int(round((h - th) / 2.))
    j = int(round((w - tw) / 2.))
    return video[..., i:(i + th), j:(j + tw)]

# Normalize using mean and standard deviation
def normalize_base(video, mean, std):
    shape = (-1,) + (1,) * (video.dim() - 1)
    mean = torch.as_tensor(mean).reshape(shape)
    std = torch.as_tensor(std).reshape(shape)
    return (video - mean) / std
```

We will use torchvision.io method to read a video file and show shape and some other useful information about the video we use as a source:

```
import torchvision.io
video_file = 'media/surfing_cutback.mp4'
video, audio, info = torchvision.io.read_video(video_file)
shape = video.shape
print(f'frames {shape[0]}, size {shape[1]} {shape[2]}\n{info}')
```

```
Output:
frames 255, size 720 1280 {'video_fps': 29.97002997002997, 'audio_fps':
48000}
```

As you can see, the original video has 255 frames and 720p resolution, which is relatively large for our model. As discussed earlier, we need to normalize the video before submitting it to the model; this is the time to call the normalizing methods we defined earlier:

```
video = normalize(video)
video = resize(video,(128, 171))
video = crop(video,(112, 112))
video = normalize_base(video, mean=mean, std=std)
shape = video.shape
print(f'frames {shape[0]}, size {shape[1]} {shape[2]}')

Output:
frames 3, size 255 112
```

Much better! After normalization, the video is much smaller, and you can experiment with the number of frames (authors of the original model used in PyTorch mention various clip lengths of 8, 16, 32, 40, and 48 frame clips; in our experiment we only use 3 frames for inference). If you have a GPU-enabled device with CUDA, you can accelerate the process by moving both model and video tensor to the CUDA-enabled device:

```
# Make use of accelerated CUDA if available
if torch.cuda.is_available():
    model.cuda()
    video = video.cuda()
```

Now comes the magic of applying our pretrained model and making it process an actual surfer video. Note that in the first line, the model expects a tensor with one extra dimension, so in the call to the model, we apply unsqueeze method to the video to satisfy the model's requirements. The result is an array of scored classes (activities). In the second line, we use argmax() function to select the best matching value and assign it to prediction. We then print the best score, which is the number of the class in the list of activities of Kinetics dataset. This may take some time if you have a CPU only, depending on your environment, so be patient:

```
# Score the video (takes some time!)
score = model(video.unsqueeze(0))
# Get prediction with max score
prediction = score.argmax()
print(prediction)

Output:
tensor(337)
```

The resulting index is not very meaningful, so let's convert it to the actual class name it represents, by using our utility script kinetics.classes(). That returns a list of classes in the Kinetics dataset which makes it very easy to map the result of the prediction to the name of the activity classes[prediction.item()]:

```
from utils.kinetics import kinetics
classes = kinetics.classes()
print(classes[prediction.item()])

Output:
surfing water
```

And it turns out to be "surfing water", the class Kinetics model was trained with to detect a surfing action. Our predicted result is correct! In this example we used a custom video file from a consumer grade 720p resolution video camera. We used a PyTorch pretrained model trained on Kinetics dataset for video classification of 400 activities, of which more than a hundred are sports related. We normalized the video and classified an activity correctly on the video, by returning the best score (score. argmax) and mapping the numerical result to the index in the list of activities in Kinetics dataset.

Loading Videos for Classification

PyTorch includes a number of modules simplifying video classification. In the previous project, you already explored an introduction to video classification, based on a pretrained models, included in torchvision. We also used torchvision.io.read_video method to load videos in a convenient structure of tensors that include video frames, audio, and relevant video information. In the following project, we'll take it further and will do some practical video loading and model training, as well as transfer learning for video classification.

PROJECT 9-2: LOADING VIDEOS FOR CLASSIFIER TRAINING

In this project I'll show you how to use video dataset modules, such as Kinetics400 and DataLoader to visualize videos and prepare them for training. Kinetics folder structure follows a common convention that includes train/test/validation folders and videos split into classes of actions we need to recognize. Since datasets, such as Kinetics, are based on indexed online videos, there're many scripts out there that simplify loading videos for training and structuring them in folders. For now, we'll define the base directory of our dataset:

```
base_dir = Path('data/kinetics400/')
data_dir = base_dir/'dataset'
```

Conveniently, as part of torchvision.datasets, PyTorch includes Kinetics400 dataset that serves as a cookie cutter for our project. Internally, video datasets use VideoClips object to store video clips data:

```
data = torchvision.datasets.Kinetics400(
            data_dir/'train',
            frames_per_clip=32,
            step_between_clips=1,
            frame_rate=None,
            extensions=('mp4',),
            num_workers=0
        )
```

▊ **Note** Although you can and should take advantage of the multiprocessing nature of datasets, especially in the production environment, on some systems you may get an error; num_workers = 0 makes sure you use dataset single threaded.

According to this constructor earlier, each video clip loaded with our dataset should be a 4D tensor with the shape (frames, height, width, channels); in our case 32 frames, RGB video, note that Kinetics doesn't require all clips to be of the same height/width:

```
print((data[0][0]).shape)
Output:
torch.Size([32, 226, 400, 3])
```

Visualizing Dataset

Sometimes, it may be handy to visualize the entire dataset catalog as a table, summarizing the number of frames. The helper function to_dataframe loads the entire video catalog into Pandas DataFrame and displays the content:

```
from utils.video_classification.helpers import to_dataframe
```

```
to_dataframe(data)
```

Let's say we want to display the size of a video in the dataset:

```
VIDEO_NUMBER = 130
video_table = to_dataframe(data)
video_info = video_table['filepath'][VIDEO_NUMBER]
```

With notebook IPython.display video helper, we can also show the video embedded in the notebook (Figure 9-5), but keep in mind that setting embed=True while displaying the video may significantly increase the size of your notebook:

```
from IPython.display import Video
from IPython.core.interactiveshell import InteractiveShell
InteractiveShell.ast_node_interactivity = "all"

Video(video_info, width=400, embed=False)
```

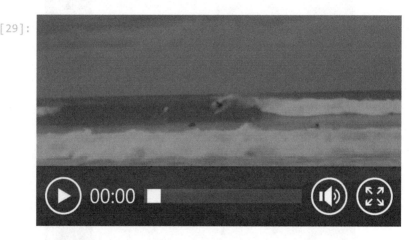

[29]:

Figure 9-5. Displaying embedded video in notebooks

So instead of embedding the video, it may be sufficient to just visualize the
first and last frames (Figure 9-6):

```
def show_clip_start_end(f):
    last = len(f)
    plt.imshow(f[0])
    plt.title(f'frame: 1')
    plt.axis('off')
    plt.show()
    plt.imshow(f[last-1])
    plt.title(f'frame: {last}')
    plt.axis('off')
    plt.show()

show_clip_start_end(data[0][0])
```

frame: 1

frame: 32

Figure 9-6. Loading and visualizing video frames

Video Normalization

As with most of the data, before training our model, video needs to be normalized for video classification models included in torchvision. This involves
getting image data in the range [0,1] and normalizing with standard deviation
and the mean provided with the model:

```
t = torchvision.transforms.Compose([
      T.ToFloatTensorInZeroOne(),
      T.Resize((128, 171)),
      T.RandomHorizontalFlip(),
      T.Normalize(mean=[0.43216, 0.394666, 0.37645],
                        std=[0.22803, 0.22145, 0.216989]),
      T.RandomCrop((112, 112))
   ])
```

This transformation uses Compose method from PyTorch, which combines several transformations including converting the video to a float tensor, resize, cropping, and normalizing with a mean and standard deviation. Note that in this transformation we also apply an augmentation function T.RandomHorizontalFlip() which provides a random horizontal flip to a frame. Once we've defined the transform, we can pass it to the Kinetics400 dataset:

```
train_data = torchvision.datasets.Kinetics400(
        data_dir/'train',
        frames_per_clip=32,
        step_between_clips=1,
        frame_rate=None,
        transform=t,
        extensions=('mp4',),
        num_workers=0
    )
```

DataLoader class in PyTorch provides many useful features and makes it easy to use from Python, including iterable datasets, automatic batching, memory pinning, sampling, and data loading order customization. Learning rate, as a hyperparameter for training neural networks, is important: if you make learning rate too small, the model will likely converge too slowly.

Mysterious constant   The so-called Karpathy constant defines the best learning rate for Adam as 3e-4. The author of the famous tweet in data science, Andrej himself in the response to his own tweet, says that this was a joke. Nevertheless, the constant made it to Urban Dictionary and many data science blogs.

As an illustration (Figure 9-7), notice that by making learning rate too large for gradient descent, the model will never reach its minimum.

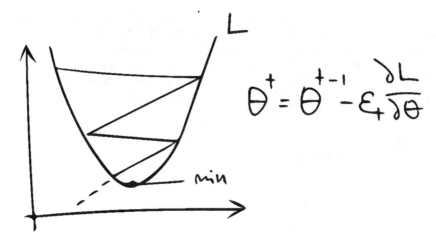

$$\theta^+ = \theta^{+-1} - \varepsilon_+ \frac{\delta L}{\delta \theta}$$

Figure 9-7. Large learning rate may miss the gradient descent minimum and the model will fail to converge. L is the loss function and et is the learning rate and θ is weights (parameters)

To deal with this problem, many frameworks, including fastai and PyTorch, now include learning rate finder module.

In case of video action recognition, with the size of the data and large differences in training times for video data, it is recommended to use a proper learning rate, which is often in the middle of the descending loss curve.

Training the Model

Training the model for video action recognition in PyTorch follows the same principles as for image classifier, but since video classification functionality is relatively new in PyTorch, it's worth including a small example in this chapter.

PROJECT 9-3: VIDEO RECOGNITION MODEL TRAINING

To start, let's create two datasets, for training and validation, based on built-in Kinetics object. The idea here is to take advantage of built-in objects that PyTorch offers for simplicity. I use the same normalizing video transformation T, already used in previous examples. On some systems you can get a significant speed improvement if you set num_workers > 0, but on my system I had to be conservative, so I keep it at zero (basically, it means don't take advantage of parallelization):

```
train_data = torchvision.datasets.Kinetics400(
          data_dir/'train',
          frames_per_clip=32,
          step_between_clips=1,
```

```
            frame_rate=None,
            transform=t,
            extensions=('mp4',),
            num_workers=0
    )

valid_data = torchvision.datasets.Kinetics400(
            data_dir/'valid',
            frames_per_clip=32,
            step_between_clips=1,
            frame_rate=None,
            transform=t,
            extensions=('mp4',),
            num_workers=0
    )
```

PyTorch allows using familiar DataLoaders with video data, and for video data PyTorch includes `VideoClips` class used for enumerating clips in the video and also sampling clips in the video while loading. `FirstClipSampler` in the following example used `video_clips` property from the dataset to sample a specified number of clips in the video:

```
train_sampler = FirstClipSampler(train_data.video_clips, 2)
train_dl = DataLoader(train_data,
                    batch_size=4,
                    sampler=train_sampler,
                    collate_fn=collate_fn,
                    pin_memory=True)
valid_sampler = FirstClipSampler(valid_data.video_clips, 2)
valid_dl = DataLoader(valid_data,
                    batch_size=4,
                    sampler=valid_sampler,
                    collate_fn=collate_fn, pin_memory=True)
```

Loading and renormalizing video data can take a really long time, so you may want to save the normalized dataset in cache directory:

```
import os
cache_dir = data_dir/'.cache'
if not os.path.exists(cache_dir):
    os.makedirs(cache_dir)
cache_dir.ls()

torch.save(train_data, f'{cache_dir}/train')
torch.save(valid_data, f'{cache_dir}/valid')
```

Next, you initialize the model with hyperparameters, including the learning rate obtained earlier. Note that since we'll be training the model, we instantiate it without weights (`pretrained=False` or omitted):

```
model = models.video.r2plus1d_18()
```

Next, we can train the model; in the following example, I chose 10 epochs:

```
for epoch in range(10):
    train_one_epoch(model,
                    criterion,
                    optim,
                    lr_scheduler,
                    train_dl, device,
                    epoch, print_freq=100)
    evaluate(model,
             criterion,
             valid_dl,
             device)
```

You can also save the model weights once it's trained:

```
SAVED_MODEL_PATH = './videoresnet_action.pth'
torch.save(model.state_dict(), SAVED_MODEL_PATH)
```

In the previous sections, we used a pretrained model; in this example, we trained a model and saved the weights. This resulting model can now be loaded and used to predict actions from a video. As a data scientist working with video recognition, you may need this to improve accuracy, or add new activities. Kinect is a large dataset, but it doesn't cover everything, so you may need to retrain your models based on activities you need to classify.

Summary

In this chapter we covered practical methods and tools for video action recognition and classification. We discussed data structures for loading, normalizing, and storing videos; datasets for sport action classification, such as Kinetics; and deep learning models. Using readily available pretrained models, we can classify hundreds of sport actions and train the models to recognize new activities. For a sport data scientist, this chapter provides practical examples for deep learning, movement analysis, and action recognition on any video.

Although video action recognition is becoming more usable today, and made progress in thousands of classifications, we are still far from the goals of generalized action recognition. That means, as a sport data scientist, you are still left with a lot of work to apply video recognition in the field. Is this the right time to make video action recognition a part of your toolbox? With practical examples and notebooks accompanying this chapter, I think that this is the right time for coaches and sport scientists to start using these methods in everyday sport data science. Video action recognition requires a lot of compute power, so you may want to consider referring to chapters of this book that describe using the cloud and automating model training.

Reinforcement Learning in Sports

Whenever I get hurt, just get me chocolate and ice cream.

—Lindsey Vonn, US Ski Team Racer

Introduction

Sports is deeply connected with coaching and learning by trying (Figure 10-1): every greatest athlete made first steps in sports by trying and taking guidance from parents, coaches, inspiration from peers, and other athletes. When I learned to ski as a kid, every word I heard from my coach had a deep impact; it made total sense, even the smallest hint or observation after a run. A bunch of skiing kids, we played, we tried again and again, and I could always tell from the coach's voice when I did something right or when I screwed.

© Kevin Ashley 2020

K. Ashley, *Applied Machine Learning for Health and Fitness*,
https://doi.org/10.1007/978-1-4842-5772-2_10

Figure 10-1. Reinforcement learning models have many similarities to coaching

Reinforcement learning (RL) is learning what to do by trying and taking the next action based on the reward (Figure 10-2). Read that sentence again, and you'll see that it makes total sense for sports: without trial and feedback, we keep running in circles. Unlike other methods in machine learning that rely on models, reinforcement learning starts with an *agent*, seeking to achieve his goal by interactively trying and receiving a reward signal for his actions. The *environment* receives actions and emits rewards, numerical values indicating how good agent's actions were.

Figure 10-2. Reinforcement learning environment, states, action, and reward

In this chapter we'll dig into reinforcement learning and its applications in sports using practical examples. You'll be surprised how many sports we can apply reinforcement learning to, right out of the box! In this chapter I'll cover examples in skateboarding, gymnastics, surfing, snowboarding, and skiing, to name a few.

For a sport scientist, reinforcement learning feels "just right" as a methodology. The idea sounds immediately understandable and fitting the game. Even the tools in reinforcement learning sound like sports: with names and terms like gym, coach, actor-critic, and *training*, any sport data scientist can find many useful applications in reinforcement learning methods. Let's explore how good this deep learning approach is in practice and learn some tips and tricks along the way.

Figure 10-3. It may sound like Groundhog Day movie, taking millions of iterations to train a human skill

---

It may sound like "Groundhog Day" movie: "It's the same thing every day", it takes 60 million samples to train a human skill with existing RL algorithms, such as DeepMimic, that's 2 days on 8 core machine (Figure 10-3)!

Tools

Reinforcement learning is an area of unsupervised learning that emerged from dynamic programming, game theory, theory of control, and models based on Markov decision process. While there're many methods and algorithms in reinforcement learning: Deep Q Networks (DQN), Deep Deterministic Policy Gradients (DDPG), Generative Adversarial Limitation Learning (GALL), and various implementations of actor-critic (A2C) are the most popular. Actor-critic method especially intuitively sounds very close to any coach's heart and athlete-coach model in particular. In sport coaching we have two neural networks: the coach that measures how good the athlete is and provides the feedback and the athlete performing the action.

The field of RL tools is developing rapidly: OpenAI Gym has emerged as a classic framework for reinforcement learning benchmarking and research, used by many data scientists, and I'll cover it in this book. OpenAI has been created as a simulation environment, a playground to unfold the "scene" of the action and train models. It remains open for research and at what algorithms should be employed to solve actual problems, with main areas being games, 2D physics problems on Box2D, classic control physics, 3D physics, and robotics, such as PyBullet and text.

Throughout this book I used generalized reinforcement learning frameworks like scikit-learn, Keras, and PyTorch; all of them can also be applied to solving reinforcement learning in the OpenAI Gym. But to simplify coding, in this chapter I'll introduce you to more specialized libraries specifically for reinforcement learning, such as the reinforcement learning baselines library. We'll also use some of the physics libraries, such as PyBullet, especially when working with complex joints and humanoid models. Production-level tools are also being developed: as an example, take a look at Microsoft Bons.ai that is also used in manufacturing.

PROJECT 10-1: APPLYING REINFORCEMENT LEARNING IN SKATEBOARDING

Let's go back to the problem we briefly touched on at the end of Chapter 2, the skateboarder pumping to get up on the wall of a half-pipe. To remind you, we are trying to solve the problem that is immediately applicable in multiple sports: riding half-pipe in skateboarding, snowboarding, and skiing when athletes need to generate momentum to gain gravitational potential energy that is further converted to kinetic energy for the next trick or dropping in. It is also common in surfing, when surfers ride the wave in a sine motion, by riding the face of the wave, bottom turning to get on the top of the wave, and dropping in again, whether with a cutback or other maneuvers that convert potential energy back to kinetic again.

Action and Observation Spaces

I don't think so much about it anymore, I just do it.

— Bethany Hamilton, Soul Surfer

OpenAI Gym has several environments that are applicable to many sports with half-pipe elements, like skateboarding, snowboarding, and freestyle skiing, and wave sports, like surfing. Historically, reinforcement learning frameworks include several classical tests for training models, from mechanics and control, for example, a test called "mountain car." The idea of this test is to train the model to drive the car back and forth to gain enough momentum to get on top of a steep hill. Every gym environment comes with an action space, represented by continuous or discrete actions: for a mountain car those actions include driving the car forward (right) or in reverse (left) (Figure 10-4).

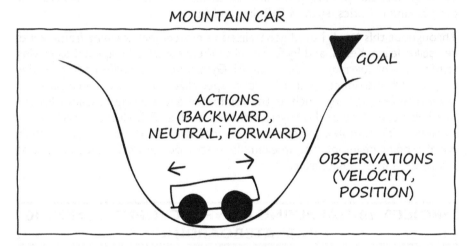

Figure 10-4. Mountain car test in reinforcement learning

In OpenAI Gym you can choose between two environments: MountainCar and MountainCarContinuous. The names indicate that one is focused on discrete action space, and another one is using continuous actions. Observations are defined by position and velocity, resulting from the actions. You can check what format those spaces are in Python by using action_space and observation_space attributes; for our continuous environment, those spaces are:

```
print(env.action_space)
print(env.observation_space)

Output:
Box(1,)
Box(2,)
```

The Box type means that we are dealing with a continuous space. The Discrete type for an action space attribute means that the agent has a discrete number of options to move. Many reinforcement learning problems include discrete actions, but you can also use an environment with a continuous space.

When selecting a proper algorithm for your environment, you should take into account that some reinforcement learning algorithms, for example, actor-critic, can solve both discrete and continuous problems, Deep Q Networks are suitable for discrete problems, and policy gradients are shown to work best with continuous spaces.

Visualizing Sample Motion

Simulation is key to movement analysis; most people are visual learners, and as much as I can describe action with motion and formulas to you, bringing a visual simulation often helps us learn better. Most reinforcement learning frameworks include tools for visualizing environments and often provide image and video rendering. To quickly visualize an environment in OpenAI Gym, you can have the agent take random actions by calling env.action_ space.sample() in a loop. Our model is not trained yet to do anything meaningful, but this method will give you an idea of how the simulation works.

In the notebook for this chapter, I'm using a couple of helper methods to simplify rendering the environment, included as a Python module—plot_ init method creates a plot and sets the initial image and plot_next renders the next iteration, by calling env.render(mode='rgb_array'):

```
import gym
import random
import numpy as np
import matplotlib
import matplotlib.pyplot as plt
from IPython import display
%matplotlib inline

def plot_init(env):
    plt.figure(figsize=(9,9))
    return plt.imshow(env.render(mode='rgb_array')) # only call this once

def plot_next(img, env):
    img.set_data(env.render(mode='rgb_array')) # just update the data
    display.display(plt.gcf())
    display.clear_output(wait=True)
```

Once these helpers are defined and included, it's easy to visualize random agent movements in the environment (Figure 10-5):

```
env = gym.make('MountainCarContinuous-v0')
env.reset()
img = plot_init(env)
for _ in range(100):
```

```
    plot_next(img, env)
    action = env.action_space.sample()
    env.step(action)
env.close()
```

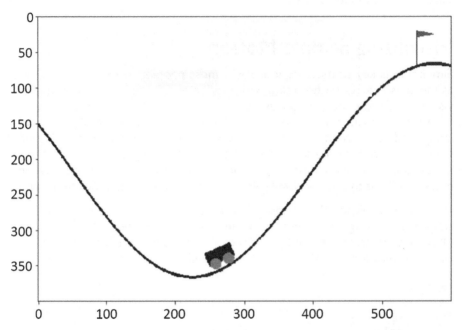

Figure 10-5. Classic control environments in reinforcement learning, such as OpenAI Gym MountainCar, can be used for sports like skateboarding, snowboarding, skiing, and surfing

Training the Model

Every loss teaches you something. The quicker you learn from the losses, then forget about the actual losing, the better off you will be. And do it fast!

—Maria Sharapova, Tennis Player, Unstoppable

Let's see how we can train our skateboarder to pump to climb to the top of the half-pipe. I'll be using actor-critic method. Let's imagine that my skateboarder's coach is providing the feedback to the athlete. In the following code, in just a few lines of code I import actor-critic (A2C) model from RL Baselines library and instantiate it with the gym environment, in our case MountainCarContinuous-v0:

```
import gym
from stable_baselines import A2C

model = A2C('MlpPolicy', 'MountainCarContinuous-v0', verbose=1, tensorboard_
log="./a2c_mcc_tensorboard/")

model.learn(total_timesteps=25000)
model.save("a2c_mcc")
```

In actor-critic model, there're two networks in play: one for the actor (our skateboarder) and one for the critic (or policy). The policy network is typically based on multilevel perceptron (we discussed it in Chapter 4, "Neural Networks") or a convolution network. The library is integrated with TensorBoard, a great tool for machine learning visualization and experimentation (Figure 10-6), so we can monitor many aspects of training and even embed in the notebook by using %tensorboard magic:

```
import tensorflow as tf
import datetime
# Load tensorboard notebook extension
%load_ext tensorboard
# Start monitoring
%tensorboard --logdir ./logs/a2c_mcc_tensorboard
```

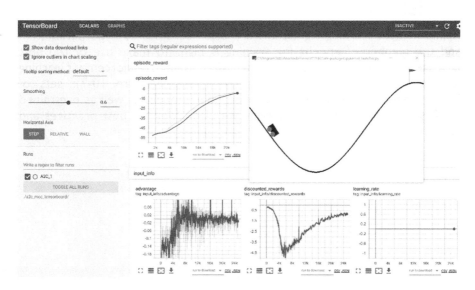

Figure 10-6. Using TensorBoard to monitor reinforcement learning

Once trained, the model is saved in a2c_mcc file and can be reused. To show how it works, simply load the model and use model.predict method to predict the movement:

```
import gym
from stable_baselines import A2C

model = A2C.load("a2c_mcc")

env = gym.make('MountainCarContinuous-v0')
obs = env.reset()
img = plot_init(env)
while True:
    action, _states = model.predict(obs)
    obs, rewards, dones, info = env.step(action)
    plot_next(img, env)
```

To summarize, we just trained our model, visualizing training progress and results with TensorBoard; our trained model was saved in a file. To use the model, we loaded it again and demonstrated the result of model's predictions in a simulated environment that OpenAI Gym provides.

Model Zoo

Reinforcement learning algorithms are often sensitive to hyperparameters and the number of epochs for training is high. To make use of the best practices and pretrained models (or behavior cloning), reinforcement learning libraries like RL Baselines include model zoos (Figure 10-7): sets of hyperparameters you can use in training your models, as well as pretrained models. As practical sport scientists, we often look for best practices and solutions that just work.

Figure 10-7. Using pretrained models and tuned parameters can significantly reduce training timeline

For example, to make use of expertly created hyperparameters for our skateboarding problem with a little Python script conveniently called enjoy.py:

```
python enjoy.py --algo a2c --env MountainCarContinuous-v0
```

The fact that some environments are based on continuous and some use discrete actions is the first hint on which algorithm to select. Working with reinforcement learning, you may soon realize that different algorithms work better for different problems. In case you wonder how hyperparameters are tuned, feel free to check out libraries like Optuna, which can make a great addition to your data scientist toolbox and help you tune the training. Fortunately, ready-to-use pretrained models are often available as part of model zoos and as part of many libraries.

PROJECT 10-2: REINFORCEMENT LEARNING IN GYMNASTICS

In gymnastics one of the fundamental tricks is the swing. Let's use reinforcement learning to solve a classic control problem: keeping the gymnast upright during the high swing. Imagine that the goal for our gymnast is to reach the top swing position and stay there with a minimal effort. The motion of the swing can be described by the pendulum model. From Newton's second law, the force F acting on the gymnast is equal to mass m times acceleration a:

$$F = ma$$

The angle of rotation (Figure 10-8) is defined as θ, mass of the gymnast is m, the radius to the center of mass is R, and FT is the tension force on the arms. The position of the center of mass of the athlete can be expressed as:

$$x = R\sin\theta\left(t\right)$$

$$y = R\cos\theta\left(t\right)$$

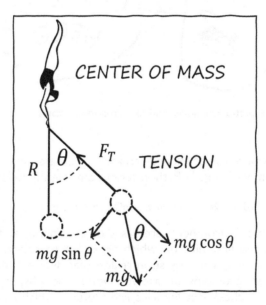

Figure 10-8. Physics of high swing in gymnastics and pendulum model

The angular velocity relative to the center of mass is $R\dfrac{d\theta}{dt}$ and acceleration is $R\dfrac{d^2\theta}{dt^2}$. The resulting force is equal to:

$$F = ma = F_T - mg\cos\theta$$

Since the gymnast is performing a circular movement with radius R and an angular velocity ω, the resulting acceleration is centripetal directed toward the center of rotation:

$$F_T = mg\cos\theta + m\frac{\omega^2}{R}$$

With this definition of the physical model for pendulum, and applications in gymnastics, such as high swing, let's see how reinforcement learning can help simulating one of the fundamental tricks in gymnastics.

Pendulum Model

In OpenAI Gym among classic control problems that reinforcement learning can solve, there's an environment for a pendulum. Let's instantiate and plot this environment in the notebook with some sample actions, visualizing a random simulation:

```
env = gym.make('Pendulum-v0')
env.reset()
img = plot_init(env)
for _ in range(100):
    plot_next(img, env)
    action = env.action_space.sample()
    env.step(action)
env.close()
```

Let's examine the pendulum environment closely: the action space is continuous, a real number, and the observation space is defined by values: cos $\theta : [-1.0, 1.0]$, sin $\theta : [-1.0, 1.0]$, where $\theta : [-\pi, \pi]$. The goal is to remain at zero angle, with the least effort. The earlier simulation shows a pendulum moving in a circular motion, but randomly. In order to perform an action simulating a high swing in gymnastics with the pendulum model, we can train the model to do it or load a pretrained model that knows how to do the trick. Fortunately, reinforcement learning baselines library includes many pretrained environments using various algorithms. Let's load a pretrained actor-critic model that solves the high swing trick:

```
import gym
from utils.gym import gymplot
from stable_baselines import A2C

model = A2C.load("data/rl/Pendulum-v0.pkl")
```

```
env = gym.make('Pendulum-v0')
obs = env.reset()
img = gymplot.plot_init(env)
for t in range(200):
    action, _states = model.predict(obs)
    obs, rewards, done, info = env.step(action)
    gymplot.plot_next(img, env)
    if done:
        print(f"Episode finished after {t+1} steps")
        break
```

When you run this part of the notebook, the simulation shows that the agent swings a few times, then reaches the z ero angle, and remains there, which is the condition the pendulum environment is targeting.

Humanoid Models

Reinforcement learning scenarios introduced in the beginning of this chapter are great easy examples that help you get started with the concept and get familiar with the tools. Pendulum and MountainCar environments are classic dynamics and control scenarios with some applications introduced in the beginning of this chapter, including half-pipe sports such as skateboarding, snowboarding, skiing, and surfing, as well as high-bar swing in gymnastics.

Can methods we discussed earlier be used for motion analysis and simulation of full humanoid models?

Figure 10-9. Simulation of a humanoid backflip

Joints and Action Spaces

Humanoid models bring us to far more realistic simulations of the humans based on biomechanics, set in a more complex environment, that requires tracking multiple joints in the human body (Figure 10-9). You will notice many interconnections between robotics, animation, motion capture, gaming, and biomechanics while playing with humanoid models and reinforcement learning. The same models that robotics engineers use in Robot Operating System in a format called URDF (Unified Robotic Description Format) can be used in simulations of humanoid models. The following code uses a humanoid model in humanoid.urdf file from PyBullet physics library to list the joints in the humanoid model:

```
import pybullet as p

p.connect(p.DIRECT)
human = p.loadURDF("humanoid/humanoid.urdf")
p.getNumBodies()
for i in range (p.getNumJoints(human)):
    jointInfo=p.getJointInfo(human,i)
    print("joint",jointInfo[0],"name=",jointInfo[1].decode('ascii'))
joint 0 name= root
joint 1 name= chest
joint 2 name= neck
joint 3 name= right_shoulder
joint 4 name= right_elbow
joint 5 name= right_wrist
joint 6 name= left_shoulder
joint 7 name= left_elbow
joint 8 name= left_wrist
joint 9 name= right_hip
joint 10 name= right_knee
joint 11 name= right_ankle
joint 12 name= left_hip
joint 13 name= left_knee
joint 14 name= left_ankle
```

Visualizing our humanoid model is also easy in PyBullet (Figure 10-10); we can simply plot the humanoid model loaded earlier into the physics environment with a simple code:

```
import pybullet as p
import pybullet_data as pd
import matplotlib
import matplotlib.pyplot as plt
import numpy as np
from utils.gym.gymbullet import get_camera_image
%matplotlib inline
```

```
p.connect(p.DIRECT)
p.resetSimulation()
p.setAdditionalSearchPath(pd.getDataPath())
p.setGravity(0,0,-9.8)
p.loadSDF("stadium.sdf")
p.loadMJCF("mjcf/humanoid_fixed.xml")

img = get_camera_image()
plt.imshow(img[2])
plt.draw()
plt.show()
p.disconnect()
```

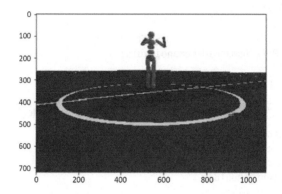

Figure 10-10. Rendering humanoid model in a physics environment

We can also create an animated image, making the camera rotate around the humanoid:

```
from utils.gym.gymbullet import animated_humanoid
import pybullet as p
import pybullet_data as pd
import matplotlib.pyplot as plt
from matplotlib import pylab
from IPython.display import Image
%matplotlib inline

p.connect(p.DIRECT)
p.setAdditionalSearchPath(pd.getDataPath())
p.resetSimulation()

plane = p.loadSDF("stadium.sdf")
human = p.loadMJCF("mjcf/humanoid_fixed.xml")
```

```
ANIMATION = "data/testing/humanoid.png"
animated_humanoid(ANIMATION)
Image(filename=ANIMATION)
```

Multiple joints in humanoid models increase complexity and number of variables in the action state. For most of the classic control scenarios like pendulum or MountainCar, action and observation spaces contain just a few variables. For humanoid models, we have more variables to track: for example, 17 for action space joints and 44 for observation space in the case of PyBullet humanoid:

```
import gym
from utils.gym import gymplot
from utils.gym import gymbullet

env,agent = gymbullet.load_humanoid()
print(f'actions: {env.action_space}')
print(f'observations: {env.observation_space}')
Output:
actions: Box(17,)
observations: Box(44,)
```

In the previous chapters, we discussed body pose estimation and humanoid models with multiple joints used in machine learning. Some practical implementations of training models with reinforcement learning have also been added to PyBullet library from an excellent research paper on DeepMimic, applying a synthesis of mocap data with reinforcement learning.

Human Motion Capture

For humanoid models training with reinforcement learning, we need to be able to have a recording of human motion or mocap. There're many ways to record human movement: video is the obvious choice. But one issue arises quickly with any video because it often includes occluded areas, for example, if our camera is facing the right part of the body, it's hard to see the what's happening with the left part. From the previous chapters, you learned some of the methods for 2D and 3D body pose reconstruction. In the professional game animation however, instead of inferred data, motion capture often uses more precise ground truth methods to capture precise movements of the human body. This empirical data captured with multiple sensor devices, such as inertial measurement units (discussed earlier in Part I), doesn't suffer from occlusion and provides a very high-frequency data of human movement (Figure 10-11). For example, an inexpensive accelerometer and gyroscope provide a 3D vector of acceleration and rotation at a rate of 100 samples per second. With professional full 3D body sensor suits like Xsens, combining 17 synchronized and calibrated sensors positioned to capture motion of every major joint, the task of collecting high-quality ground truth data becomes easy.

Figure 10-11. Motion capture with IMU sensors for ground truth

Collecting motion capture data has been an extensive area of research, for example, Carnegie Mellon University maintains a well-known motion capture database. The most common formats of motion capture file are BVH, C3D, and FBX often used in visualization and modeling tools such as Blender, Maya, and Unity (Figure 10-12).

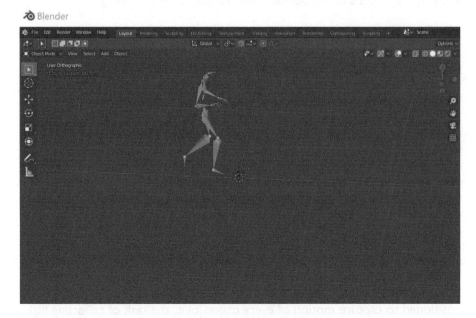

Figure 10-12. Loading a basketball dribbling mocap into Blender

Reinforcement Learning for Humanoids

By combining a motion-imitation objective with a task objective, we can train characters that react intelligently.

—Xue Bin Ping et al., DeepMimic

What reinforcement learning is truly great at is the ability to train a humanoid model to perform a realistic action. Without reinforcement learning, models may be based on kinematics or physics, but getting them to perform human-like actions and skills (Figure 10-13) without reinforcement learning has not been very successful: many efforts ended up getting unnatural looking movements. Classical mechanics serves us well, providing a framework in which gravity and basic joint mechanics control the model, but firing muscles to move in a certain (and with a certain *style*) requires a human actor for motion insights. To illustrate, let's load an *untrained* humanoid model into gym and let it randomly move; the result is likely our ragdoll falling to the ground:

```
import gym
from utils.gym import gymplot
from utils.gym import gymbullet

env,agent = gymbullet.load_humanoid(trained=False)
env.reset()
img = gymplot.plot_init(env)
for _ in range(100):
    gymplot.plot_next(img, env)
    action = env.action_space.sample()
    env.step(action)
env.close()
```

Figure 10-13. Humans and robots learning to walk

We created a connected humanoid with joints and placed it into a physics environment under forces of gravity, but we didn't teach it the skills to walk, jump, crawl, stand up, or make any actions that a human learns throughout his life. Note that in the previous example, I used sample() method to simulate random actions. In the following code sample, instead of untrained model, I load a trained model, specified by trained=True flag, and instead of the sample method, I'm using step() method to make it act using predicted movements:

```python
import gym
from utils.gym import gymplot
from utils.gym import gymbullet

env,agent = gymbullet.load_humanoid(trained=True)
obs = env.reset()
img = gymplot.plot_init(env)
for i in range(0, 30000):
    action = agent.act(obs)
    obs, r, done, _ = env.step(action)
```

```
gymplot.plot_next(img,env)
if done:
    break
```

You can see that with a trained model, results are much better, the humanoid is no longer falling, but instead it is walking.

Summary

Reinforcement learning has many applications in sports, biomechanics, games, animation, and industrial applications. In this chapter we learn and apply reinforcement learning environments for various sports: skateboarding, snowboarding, surfing, gymnastics, and others, beginning with classic control problems that can be described by just a few action states and moving on to more complex areas of biomechanics describing human body with multiple connected joints, containing many degrees of freedom and multiple action states.

In the last years, reinforcement learning methods evolved that allow very close simulation of movements, such as state-of-the-art DeepMimic. We also have new tools, such as OpenAI Gym, that include simulation environments, specialized RL libraries such as gym baselines, KerasRL, as well as evolution of generic deep learning frameworks to support reinforcement learning.

Machine Learning in the Cloud

Overview

The bulk of machine learning compute tasks today happen in the data centers. As a data scientist, you may have started your research on your local computer, playing with various models, frameworks, and sets of data, but there's a good chance that when your project reaches the stage when people start using it, your experiments may need resources that only the cloud provides. The goal of this chapter is to go over some examples of how to deploy your data science project to the cloud (Figure 11-1), store data, train your models, and ultimately give your customers access to the predictions your models provide.

© Kevin Ashley 2020
K. Ashley, *Applied Machine Learning for Health and Fitness*,
https://doi.org/10.1007/978-1-4842-5772-2_11

Figure 11-1. Deploying your data science project in the cloud

Containers

At the beginning of this book when we touched on the tools used by data scientists, you remember that we discussed virtual environments. In a virtual environment, it's easy to isolate a set of tools and libraries needed by one experiment from another. Such an environment would typically include a list of libraries and dependencies, making it relatively easy to replicate a set of components for your project. Containers take virtual environments even further. With containers you can package your data science experiments and models, together with all their supporting components, and when you are ready, distribute everything to a data center, or make it a cloud service (Figure 11-2). Even if you don't use containers explicitly, most cloud services today are using them behind the scenes to isolate and package your resources.

Figure 11-2. Docker containers make packaging and distributing machine learning models easy

As a data scientist, you are familiar with various ways to get Python installed on your system. What's the docker way to get it up and running? This one liner runs Python in a new docker container on your local system:

```
docker run -it python:3.7 python
```

The magic that happens here is that even if this version of Python is not installed on your system, docker will pull it from the online repository of images, create a container, and start a Python shell in that container:

```
Unable to find image 'python:3.7' locally
3.7: Pulling from library/python
Status: Downloaded newer image for python:3.7
Python 3.7.7 (default, Mar 11 2020, 00:27:03)
[GCC 8.3.0] on linux
Type "help", "copyright", "credits" or "license" for more information.
>>>
```

Earlier in the book we used lots of Jupyter notebooks. But what if your notebook requires a set of components that you need to configure in a different environment? Conveniently, you can also run notebooks from a docker container, and the Jupyter team has provided a set of public images you can start with:

```
docker run -p 8888:8888 jupyter/scipy-notebook
```

When this container starts, you should be able to connect to a fully functional notebook through the web browser. Docker containers have become de facto standard for packaging and distributing applications. A template with a set of instructions on what defines your package is called an *image*, and you can create your own image with a *Dockerfile*. A container is essentially an instance of an image. As a practical data scientist, if you work with the cloud, you may occasionally need to wrap your model into a container.

For a practical data scientist looking for tools that simplify container development, Visual Studio Code works on most platforms: Linux, Mac, and Windows. It includes a useful Docker extension that helps working with containers, including generating and editing Docker files with syntax completion support, working with image registries, deployment, debugging, and more. The following is an example of a Dockerfile packaging PyTorch and Torchvision machine learning libraries with Python in a container (Figure 11-3).

Figure 11-3. Using VS Code extension to build and test containers

```
FROM python:3.7

# Keeps Python from generating .pyc files in the container
ENV PYTHONDONTWRITEBYTECODE 1

# Turns off buffering for easier container logging
ENV PYTHONUNBUFFERED 1
```

```
# Install pip requirements
ADD requirements.txt .
RUN python -m pip install -r requirements.txt

WORKDIR /app
ADD . /app
```

```
# During debugging, this entry point will be overridden. For more
information, refer to https://aka.ms/vscode-docker-python-debug
CMD ["python", "app.py"]
```

The container packages requirements.txt file which includes torch and torchvision dependencies and app.py file as an entry point for the container.

Notebooks in the Cloud

Running Jupyter notebooks on your local machine (Figure 11-4) is not the only way to run your data science experiments with some great (and often free to start) services available today: most major cloud vendors provide services that can get you started quickly with the notebooks. For starters some of the major notebook services available today include Microsoft Azure Notebooks, Google Colaboratory, and Amazon EMR. Beyond offering basic Jupyter notebooks, they often provide a set of tools that simplify managing machine learning workflows, loading and processing data, and other integration services, including monitoring, scaling, and source code integration.

Figure 11-4. Use cloud-based notebooks for data science experiments

We'll start by creating a free notebook using Azure Machine Learning and connecting to the workspace environment:

```
import azureml.core
from azureml.core import Workspace

workspace = Workspace.from_config()
print(workspace.name)
```

This Python snippet takes advantage of the Azure Python SDK, a set of methods that simplifies working with objects in the workspace. If you want to connect to the cloud workspace from the local Jupyter Notebook, you can simply export configuration config.json file and place it in the directory where you run your local notebook; then magically your Workspace.from_config() command will use your local configuration file to connect with your cloud environment.

Data in the Cloud

One of the most important advantages of developing your machine learning experiments in the cloud is the use of cloud-based storage. Storage in the cloud is a relatively inexpensive resource that you can easily scale, with added benefits of security, durability, and high availability across different geographical regions and accessibility anywhere in the world from many languages and platforms.

PROJECT 11-1: LOADING AND ACCESSING TABULAR SENSOR DATA IN THE CLOUD

In the previous chapters, when we discussed various ways to capture data from athletes, I mentioned IMUs (inertial measurement units) that can aggregate data from several sensors, accelerometers, gyroscopes, and magnetometers, to provide accurate high-frequency information about athletes' movements. To illustrate the use of this data in the machine learning environment in the cloud, I provided a motion capture file of a high-level skier performing slalom turns. This data was captured using high-quality mocap with an Xsens MVN full-body suit that combines multiple sensors (Figure 11-5).

In addition to storing data in the cloud, as a data scientist you are likely to spend some time parsing the file and getting it into various models for training and further processing. In the following example, we'll load an output from such a set of IMUs into the cloud-based machine learning workspace. If you are using source code notebooks, accompanying the book, data is located in data/xsens folder. The following code snippet takes a comma-separated file containing center of mass data and creates a tabular dataset, registering it in the cloud workspace.

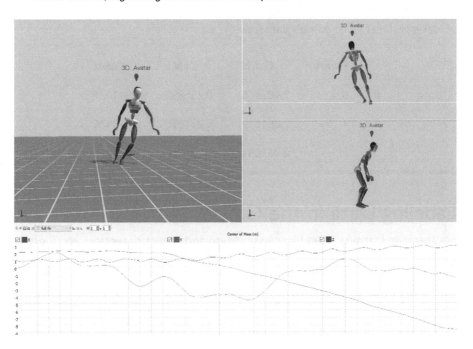

Figure 11-5. Sensor data visualization

```
import os
from azureml.core import Workspace, Datastore, Dataset

datastore = workspace.get_default_datastore()
source_dir = os.getcwd()
store_path = 'center_of_mass'

datastore.upload_files(
    files=[os.path.join(source_dir, f) for f in ['skier_center_of_mass.
    csv']],
    relative_root=source_dir,
    target_path=store_path,
    overwrite=True)
dataset
= Dataset.Tabular.from_delimited_files(path=(datastore, store_path))
dataset = dataset.register(workspace=workspace,
                           name='center_of_mass',
                           description='skier center of mass')
```

To check if our dataset loaded the file correctly, here's a quick test that displays a few lines of the loaded data:

```
# Show results
dataset.take(3).to_pandas_dataframe()
```

	Frame	CoMx	CoMy	CoMz
0	0	2.062156	-0.255855	0.914021
1	1	2.061619	-0.251858	0.910926
2	2	2.061091	-0.247852	0.907770

Labeling Data in the Cloud

For machine learning tasks like classification and object detection, you'll often need to label data for training. In the previous chapters, when we discussed deep computer vision and classification, you used a labeled dataset of different sporting activities and a pretrained model to classify activities. If you recall, we had two classes of actions: 'tennis' and 'surfing', and we trained our model with a set of preclassified images. We also used transfer learning from a pretrained model, which reduced the number of images we needed to supply for the model. The task of labeling data in the cloud often needs to be done by a distributed team, with thousands of images: the cloud comes very handy (see Figure 11-6)!

Once the data is labeled, it can be exported in COCO format (Common Objects in Context), the standard we used earlier in the book when we experimented with human body poses to store joint information. COCO data format is frequently used for object and keypoint detection, segmentation, and captioning.

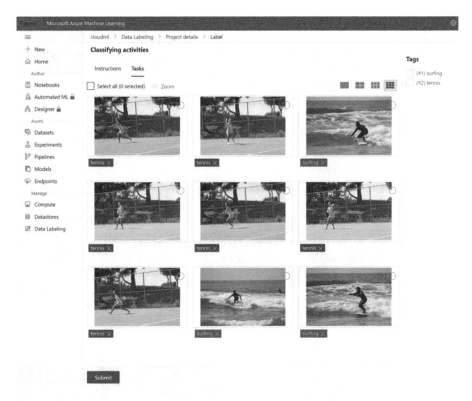

Figure 11-6. Using cloud-based labeling project for activity classification

PROJECT 11-2: TRAINING CLASSIFICATION MODEL ON THE LABELED DATASET IN THE CLOUD

Let's use the dataset we just labeled in the cloud to train our activity classification model. You probably wonder at this point, where does the labeled dataset live in the cloud and how to get access to it? The following code snippet obtains the dataset from the workspace in the cloud and then loads it as pandas dataframe:

```
from azureml.core import Dataset
from azureml.contrib.dataset import FileHandlingOption

dataset = Dataset.get_by_name(workspace, name="activities")
dataset_pd = dataset.to_pandas_dataframe(file_handling_
option=FileHandlingOption.DOWNLOAD, target_path='./download/',
overwrite_download=True)
dataset_pd)
print(dataset_pd.shape)

OUTPUT: (180, 3)
```

Note that although the labeled dataset contains URLs to images, you also have ability to download image files locally, by using `FileHandlingOption.DOWNLOAD`. Once the data is labeled and exported, you can visualize it using standard Python libraries (Figure 11-7). Note that since this is a labeled dataset, each image now includes a label of the activity ('surfing' or 'tennis'):

```python
import numpy as np
import matplotlib.pyplot as plt

w=10
h=10
fig=plt.figure(figsize=(15, 15))
columns = 2
rows = 2
for i in range(1, columns*rows +1):
    img = mpimg.imread(dataset_pd.loc[i,'image_url'])
    ax = fig.add_subplot(rows, columns, i)
    ax.title.set_text(dataset_pd.loc[i,'label'])
    ax.axis('off')
    plt.imshow(img)
plt.show()
```

['surfing']

['tennis']

['surfing']

['tennis']

Figure 11-7. Displaying labeled images in Python

In the earlier chapters, we used PyTorch's torchvision to load our dataset locally. Conveniently, our cloud-labeled dataset can be easily converted to a torchvision dataset, containing torch tensors:

```
from torchvision.transforms import functional as F

pytorch_dataset = dataset.to_torchvision()
img = pytorch_dataset[0][0]
print(type(img))

OUTPUT: <class 'torch.Tensor'>
```

Preparing for Training

Before we start training our model, we need to tell the cloud where the model is trained or specify a compute target: a VM or a compute cluster that satisfies the needs of your model, including GPU support and size. You can connect to an existing compute target, created with your workspace, or add a new one; in this case I connect to an existing compute target:

```
from azureml.core.compute import ComputeTarget, AmlCompute
from azureml.core.compute_target import ComputeTargetException

cluster_name = "compute-experiments"

compute_target = ComputeTarget(workspace=workspace, name=cluster_name)
```

You can also create a new compute target with ComputeTarget.create() method. When you configured your local computer for machine learning, in the previous chapters, you probably used Anaconda to manage virtual environments. Similarly, in the cloud you have a way to provision your compute target and the environment that your model needs; note that you can include Python packages in CondaDependencies of your environment (pretty cool, huh?):

```
from azureml.core import Environment
from azureml.core.conda_dependencies import CondaDependencies

conda_env = Environment('conda-env')
conda_env.python.conda_dependencies = CondaDependencies.create(pip_
packages=['azureml-sdk',
                                'azureml-contrib-dataset',
                                'torch','torchvision',
                                'azureml-dataprep[pandas,fuse]'])
```

Data scientists deal with many frameworks and libraries to train models, such as PyTorch, Keras, scikit-learn, TensorFlow, Chainer, and others. Most of model development falls into the same pattern: first, you specify an environment to train your model, including dependencies, parameters, and scripts that define your experiment and how the model is trained; then, the model is trained and saved or registered in the workspace. Azure ML SDK provides two useful abstractions: one that wraps our experiments in the Experiment object and another one, called Estimator, that simplifies model training. In the following code snippet, I create an experiment and an estimator with a script named `train.py` we'll discuss in the next section:

```
import os
from azureml.train.estimator import Estimator
from azureml.core import Experiment

experiment = Experiment(workspace=workspace, name='activity-classification')
script_folder = './activity-classification'
os.makedirs(script_folder, exist_ok=True)

estimator = Estimator(source_directory=script_folder,
                entry_script='train.py',
                inputs=[dataset.as_named_input('activities')],
                compute_target=compute_target,
                environment_definition=conda_env)
```

Model Training in the Cloud

In Chapter 6, "Deep Computer Vision," you used PyTorch to train a model to classify a sport activity. We used a local notebook to run our training, and our dataset was already labeled: all images were placed in the folders corresponding to the names of the classes: *surfing* or *tennis*.

In this cloud-based project, we will use the *activities* dataset we labeled using the cloud workflow from the previous section, and since earlier we already told the estimator where our training entry point will live, we'll place all our training code in the script `train.py`. Fortunately, we can reuse most of our model training code used for classification, making adjustments for running it in the cloud. When the training script runs in the cloud, the Run object maintains context information about our experiment environment, including input datasets we send to the model for training. You can obtain the context of the experiment by using Run.get_context() call and then get our labeled activities dataset from run.input_datasets['activities']:

```
from azureml.core import Dataset, Run
import azureml.contrib.dataset
from azureml.contrib.dataset import FileHandlingOption, LabeledDatasetTask
```

```
run = Run.get_context()
# Get input dataset by name
labeled_dataset = run.input_datasets['activities']
```

Our activities dataset contains images, and in the training script, you'll need access to these images. Images and the dataset are already stored in cloud storage; we don't need to copy them over, simply mount our dataset with Dataset.mount() command:

```
mounted_path = tempfile.mkdtemp()
# Mount dataset onto the mounted_path of a Linux-based compute
mount_context = labeled_dataset.mount(mounted_path)
mount_context.start()
print(os.listdir(mounted_path))
print (mounted_path)
```

The following load() method loads images from the labeled dataset and applies the transformation that the ResNet model requires. Remember that a pretrained model needs all images normalized in the same way. The model expects all images to be 224 pixels, with 3 RGB channels, and normalized using mean = [0.485, 0.456, 0.406] and standard deviation std = [0.229, 0.224, 0.225]. As the script loads images, it also performs normalization. We will also split the dataset between training and testing, like we did earlier when the model was trained using a local notebook:

```
def load(f, size = .2):

    t = transforms.Compose([transforms.Resize(256),
        transforms.CenterCrop(224),
        transforms.ToTensor(),
        transforms.Normalize(mean = [0.485, 0.456, 0.406],
        std = [0.229, 0.224, 0.225])])

    train = datasets.ImageFolder(f, transform=t)
    test = datasets.ImageFolder(f, transform=t)
    n = len(train)
    indices = list(range(n))
    split = int(np.floor(size * n))
    np.random.shuffle(indices)
    train_idx, test_idx = indices[split:], indices[:split]
    train_sampler = SubsetRandomSampler(train_idx)
    test_sampler = SubsetRandomSampler(test_idx)
    trainloader = torch.utils.data.DataLoader(train,sampler=train_sampler,
    batch_size=64)
    testloader = torch.utils.data.DataLoader(test, sampler=test_sampler,
    batch_size=64)
    return trainloader, testloader
```

```
trainloader, testloader = load(f, .2)
print(trainloader.dataset.classes)
images, labels = next(iter(trainloader))
```

Just like last time, we will use a pretrained ResNet model, trained with ImageNet, using transfer learning. We instruct PyTorch to avoid backpropagation by setting requires_grad to False. Then we replace the last fully connected layer with a linear classifier for two classes of our labeled dataset, *surfing* and *tennis*:

```
features = model.fc.in_features
model.fc = nn.Linear(features, len(labels))
model = model.to(device)
criterion = nn.CrossEntropyLoss()
optimizer = optim.SGD(model.parameters(), lr=0.001, momentum=0.9)
scheduler = lr_scheduler.StepLR(optimizer, step_size=7, gamma=0.1)
print_every = 100
```

The model training method is very similar to the classification example we did earlier:

```
def train_model(epochs=3):
    total_loss = 0
    i = 0
    for epoch in range(epochs):
        for inputs, labels in trainloader:
            i += 1
            inputs, labels = inputs.to(device), labels.to(device)
            optimizer.zero_grad()
            logps = model.forward(inputs)
            loss = criterion(logps, labels)
            loss.backward()
            optimizer.step()
            total_loss += loss.item()
            test_loss = 0
            accuracy = 0
            model.eval()
            with torch.no_grad():
                for inputs, labels in testloader:
                    inputs, labels = inputs.to(device), labels.to(device)
                    logps = model.forward(inputs)
                    batch_loss = criterion(logps, labels)
                    test_loss += batch_loss.item()

                    ps = torch.exp(logps)
                    top_p, top_class = ps.topk(1, dim=1)
                    equals = top_class == labels.view(*top_class.shape)
                    accuracy += torch.mean(equals.type(torch.FloatTensor)).item()
```

```
        train_losses.append(total_loss/len(trainloader))
        test_losses.append(test_loss/len(testloader))
        print(f"Epoch {epoch+1}/{epochs}.. "
                f"Train loss: {total_loss/print_every:.3f}.. "
                f"Test loss: {test_loss/len(testloader):.3f}.. "
                f"Test accuracy: {accuracy/len(testloader):.3f}")
        running_loss = 0
        model.train()
    return model
```

We can also see the model converges (Figure 11-8).

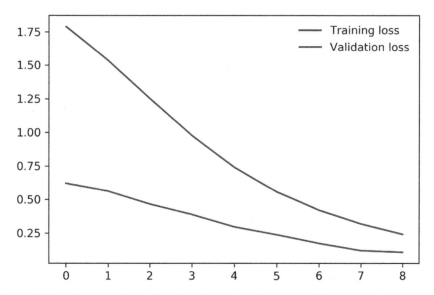

Figure 11-8. Training and validation loss for the classification model

Finally, we call the `train` method, and when the training is finished, our model is saved in the experiment instance's `./outputs` folder:

```
model = train_model(epochs=3)
print('Finished training, saving model')
os.makedirs('./outputs', exist_ok=True)
torch.save(model, os.path.join('./outputs', 'activities_classifier_model.pt'))
```

Running Experiments in the Cloud

Now, you created your compute target, experiment, training script, and estimator; you can submit your experiment and wait until the model is trained! Everything we've done in this chapter was a prelude to these two lines of code: our training script, the definition of our experiments, and cloud environment; this is where all the action takes place. The cloud interface (Figure 11-9) also provides a visual interpretation of our experiments running over time:

```
run = experiment.submit(estimator)
run.wait_for_completion(show_output=True)
```

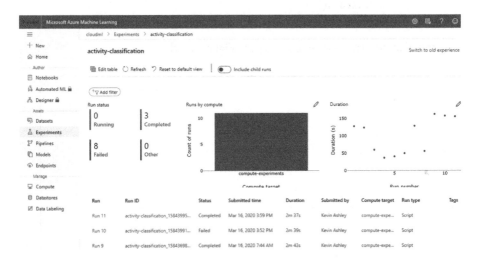

Figure 11-9. Cloud interface to visualize the progress of your experiments

---

Note The preceding command may take a long time! Your compute target is first provisioned with all required dependencies before it starts the actual training.

Model Management

After you trained your model, you can register it in the cloud. A trained model is the "brain" of your AI and can be used from an API, a web service, or any other endpoint to provide meaningful information to your customers. In this example, the model we trained to provide activity classification is registered as `activity_classification`:

```
model = run.register_model(model_name='activity_classification', model_
path='./outputs/activities_classifier_model.pt')
print(model.name, model.id, model.version, sep="\t")
```

Alternatively, you can also download the model to your local device; this model can be loaded in PyTorch:

```
run.download_file(name='outputs/activities_classifier_model.pt',
output_file_path='./models')
```

Summary

Using cloud-based machine learning methods is the natural step in bringing your experiment to your customer. In this chapter we looked at some familiar tasks, like using notebooks, loading and processing data, labeling, classification, and training your models. Everything you've done on your local computer with notebooks (and more!) you can do in the cloud. In this chapter you learned how to take a familiar task, like an image classification for different sport activities, and create a pipeline for training it in the cloud. Some new concepts include creating an environment for your experiment, including Python packages dependencies, defining a compute target to run your training, and registering your model in the cloud environment. In the next chapter, we'll look at automating these tasks and making your data science experiments available to users of your models.

Automating and Consuming Machine Learning

There's a way to do it better - find it.

—Thomas A. Edison

© Kevin Ashley 2020

K. Ashley, *Applied Machine Learning for Health and Fitness*,

https://doi.org/10.1007/978-1-4842-5772-2_12

Figure 12-1. Automating all aspects of machine learning pipeline from deployment to consumption by end users

Overview

In the previous chapter, you used some of the cloud tools to label and process our data, train the models, and register them in the cloud. As a researcher, you are likely to train quite a few models, changing your training script, experimenting with data, before making them available to your customers. This chapter is about making AI a high-quality, automated process that makes it easy to manage your code, publish models, and consume them (Figure 12-1). You'll see the term CI/CD (continuous integration/continuous delivery) many times referring to the development cycle in machine learning, and in this chapter, I'll be going over some practical examples of taking your research to the level of best practices and standards used in modern data science.

Managing Models

In the last chapter you trained a classification model that can classify images of sport activities. As a data scientist, you feel happy: your model converges, it predicts sport activities with good accuracy, and now your customer wants to use it. To make it available for your client, the model needs to be deployed, so you can give your clients something like a link to an API, and then they can use it in their own apps.

PROJECT 12-1: REGISTERING, DEPLOYING, AND CONSUMING A MODEL IN THE CLOUD

The first step in deploying your models is registering them in the workspace; this saves them in the cloud so they can be used later from your code:

```
from azureml.core.model import Model

model = Model.register(model_path = "./models",
                model_name = "activities",
                description = "Activity Classification",
                workspace = workspace)
```

Now, referencing your models becomes super easy; simply pass your workspace and model name, and in your code, you have a reference to the model:

```
model = Model(workspace, 'activities')
```

You can check the path in the cloud of the model you just deployed; note that registration automatically versions your models:

```
Model.get_model_path('activities', _workspace=workspace)
```

The location of your registered model in the workspace will become important in the next steps, because you'll need to reference this model in your scoring script's initialization, when this model is loaded by your service.

Creating a Scoring Script

Knowledge is a treasure, but practice is the key to it.

—Lao Tzu

You already created a script to train your model; another script you'll need is the scoring, or inferencing script, typically named score.py. The script is often specific to your model; in our case we use PyTorch and torchvision, but if you use a different machine learning model library, your script will have a similar structure but use methods specific to your environment and model for inference. The idea is that the script runs in the context of a web service or an API. The scoring script needs two methods: init() and run(). The first one, init(), is executed once when the container with your model is started, and loads the model and classes into a global variable:

```
def init():
    global model, classes
    model_path = os.path.join(os.getenv('AZUREML_MODEL_DIR'), 'models',
    'activities.pkl')
    model = torch.load(model_path, map_location=lambda storage, loc: storage)
    model.eval()
    classes = ['surfing','tennis']
```

The run() method is invoked each time your model is called to predict something. For the run method, since our model classifies an activity based on an image, the image needs to be decoded first from an HTTP request and then transformed according to the size, mean, and standard deviation that our model was trained with in the previous chapter:

```
def transform(image_file):
    # See PyTorch ResNet model transformations
    # our model has been trained with specific
    # size, mean, and standard deviation
    t = transforms.Compose([transforms.Resize(256),
        transforms.CenterCrop(224),
        transforms.ToTensor(),
        transforms.Normalize(mean = [0.485, 0.456, 0.406],
        std = [0.229, 0.224, 0.225])])

    image = Image.open(image_file)
    image = t(image).float()
    image = torch.tensor(image)
    image = image.unsqueeze(0)
    return image

def decode_base64_to_img(base64_string):
    base64_image = base64_string.encode('utf-8')
    decoded_img = base64.b64decode(base64_image)
    return BytesIO(decoded_img)

def run(input_data):
    image = decode_base64_to_img(json.loads(input_data)['data'])
    image = transform(image)

    output = model(image)

    softmax = nn.Softmax(dim=1)
    pred_probs = softmax(model(image)).detach().numpy()[0]
    index = torch.argmax(output, 1)

    result = json.dumps({"label": classes[index], "probability":
    str(pred_probs[index])})
    return result
```

Defining an Environment

Your inference environment defines your machine learning web service configuration. You can use either Anaconda or pip requirements file to create your environment. For example, for Anaconda, use a YAML file similar to what conda env export command generates:

```
%%writefile $script_folder/activities.yml
name: Activities-PyTorch
dependencies:
  - python=3.6.2
  - pip:
    - azureml-defaults
    - azureml-core
    - azureml-contrib-dataset
    - azureml-dataprep[pandas,fuse]
    - inference-schema[numpy-support]
    - torch
    - torchvision
    - pillow
```

Next, you create the environment by using either from_pip_requirements or from_conda_specification method:

```
from azureml.core.environment import Environment

env = Environment.from_conda_specification(name='Activities-PyTorch',file_
path=script_folder+"/activities.yml")
env.register(workspace=workspace)
```

Since you are in a big friendly cloud, there're many environments that you can easily reuse; just run this script to see a large number of environments that are available:

```
envs = Environment.list(workspace=workspace)

for env in envs:
    if env.startswith("AzureML"):
        print("Name",env)
        print("packages",
envs[env].python.conda_dependencies.serialize_to_string())
```

Deploying Models

There're many ways to deploy your models, as a local web service, Azure Kubernetes Service (AKS), Azure Container Instances (ACI), Azure Functions, and more. Each deployment type has advantages: for example, a Kubernetes-based deployment is best for production-level scalable deployments, while container instances are a fast and easy way to deploy.

In this example our model is trained with PyTorch and saved as a Python pickle .pkl file, Keras models are often saved as HDF5 .h5 files, and TensorFlow saves models as protocol buffer .pb files. Open Neural Network Exchange, or ONNX, is a promising standard that deals with interoperability of model formats and AI tools: the initialization function of your scoring script is responsible for loading the model.

It is often easy to start with a local deployment while you're developing your model; this allows you to check for any problems in your scoring and initialization script. Let's test deploying our model on the local web server, using port 8891 as an endpoint:

```python
from azureml.core.environment import Environment
from azureml.core.model import InferenceConfig, Model
from azureml.core.webservice import LocalWebservice

def deploy_locally(model_name, port):
    model = Model(workspace, model_name)
    myenv = Environment.from_conda_specification(name="env", file_
    path=script_folder+"/activities.yml")
    inference_config = InferenceConfig(entry_script=script_folder+"/score.py",
    environment=myenv)
    deployment_config = LocalWebservice.deploy_configuration(port=port)
    return Model.deploy(workspace, model_name, [model], inference_config,
    deployment_config)

service = deploy_locally('activities', 8891)
service.wait_for_deployment(True)
print(service.port)

OUTPUT:
Local webservice is running at http://localhost:8891
```

Behind the scenes, everything that happens locally will also be happening in the cloud, with our next steps. When you call Model.deploy method, your environment specification is used to build and start a docker container, the model is copied to the container, and the scoring script you created earlier is invoked at the initialization method. Let's deploy our service to the cloud now:

```python
from azureml.core.webservice import AciWebservice, Webservice
from azureml.core.model import Model

service_name = 'activity-classification'
model = Model(workspace, 'activities')
deployment_config = AciWebservice.deploy_configuration(cpu_cores = 1,
memory_gb = 1)
```

```
service = Model.deploy(workspace, service_name, [model], inference_config,
deployment_config)
service.wait_for_deployment(show_output = True)
print(service.state)
print(service.get_logs())
```

This script looks very similar to the local deployment you just did with Model.
deploy() but notice that deployment configuration is created with
AciWebservice instead of LocalWebservice, and instead of the port, you
specified cpu_cores and memory_gb as parameters to size your deployment.

In the preceding example, you loaded a single model, but what if your model
doesn't generalize well, or your API exposes multiple models? It often happens
in machine learning that you need to package many models into the same
service API. Generally, you can register multiple models and let your
initialization script load them.

CALLING MACHINE LEARNING MODELS

Figure 12-2. Providing a way for end users, apps, and services to call your model from a web
service API

Calling Your Model

The model is successfully deployed and is now ready for users to call it
(Figure 12-2). Our model accepts an image as an argument, so we need to
encode it before it is sent inside JSON, as Base64 string. In the receiving script
run() of your score.py file you created earlier, the image is decoded again and
then our model is called to predict the activity. You can test how our model
works by calling it:

```
import json
import base64

path = 'download/workspaceblobstore/activities/surfing/'

def call_inference(image_path):
    with open(image_path, 'rb') as file:
        byte_content = file.read()
    base64_bytes = base64.b64encode(byte_content)
    base64_string = base64_bytes.decode('utf-8')
    request = json.dumps({'data': base64_string })
    prediction = service.run(input_data=request)
    print(prediction)

call_inference(path+'/resize-DSC04631.JPG')

OUTPUT:
{"label": "surfing", "probability": "0.72666925"}
```

The call to our model will return the predicted activity: surfing and the probability of the prediction. Getting back to the goal we stated at the beginning of this chapter, we need to give to our customer a simple URI link they can use in a multitude of apps. To get a link to the service you just deployed, you can use `service.scoring_uri`:

```
print("Model inference URI: ", service.scoring_uri)
```

Continuous Machine Learning Delivery

> *A pile of rocks ceases to be a rock pile when somebody contemplates it with the idea of a cathedral in mind.*
>
> —Antoine de Saint-Exupéry

In the previous sections, we stepped through the process of registering a model, creating a scoring script that we used to initialize the model and provide an inferencing endpoint for the API that we published as a web service. It turns out that this process is highly repeatable in data science. As your data science team works on the models and data, improving the model's accuracy, we need to make sure we keep track of changes, issues, and new models that are deployed. It's important to follow engineering best practices to make our project continuously deliver value to customers.

Machine Learning Pipelines

Machine learning workflow needs an architecture and the level of automation that applies to all stages of AI projects: from source code management to data integration, model development, unit testing, and releasing models, to QA and production environments and monitoring and scaling the models. In the early stages of the project, you may be dealing with Jupyter notebooks, but AI projects require a solid level of automation and process management to be successful (Figure 12-3).

Figure 12-3. Machine learning pipeline for classification model training and registration

Source Code

It all starts with the source code integration: most machine learning CI/CD frameworks integrate with GitHub, DevOps, or other source control systems. Typically, as model scripts are checked into the source control by data scientists, the pipeline may be triggered to train, package, and deploy the model. The premise of continuous delivery cycle is automating this process.

Automating Model Delivery

To start with a continuous model training and delivery, frameworks such as Azure Python SDK offer some neat tools that make it easy to wrap the process into a repeatable set of steps, conveniently called a pipeline. If you are familiar with ETL processes dealing with data, then the concept should look very familiar to you. In fact, machine learning pipelines are often based on the same architecture and involve data transformation steps and repeatable processes that can be scheduled or triggered to run (Figure 12-4). For the purpose of creating the pipeline, you can reuse most scripts, such as your model training script.

CI/CD PIPELINE STEPS

Figure 12-4. Pipeline automation steps usually involve data preparation, model training, and deployment

---

PROJECT 12-2: CREATING A CONTINUOUS MODEL TRAINING PIPELINE

Runtime Environment

Before you build the pipeline, let's create an environment with dependencies we need in our model, such as PyTorch and torchvision, and runtime configuration that will be used in the pipeline. Setting `docker.enabled` in the environment also ensures that the environment supports containers:

```
environment = Environment(env_name)
environment.docker.enabled = True
environment.python.conda_dependencies = CondaDependencies.
create(pip_packages=['azureml-sdk',
                                    'azureml-contrib-dataset',
                                    'torch','torchvision',
                                    'azureml-
dataprep[pandas,fuse]'])

config = RunConfiguration()
config.target = compute_target
config.environment = environment
```

Our model is based on classification, and in the previous chapter, you created a labeled dataset as part of the workspace. We will reference that dataset in the pipeline as input for model training. The output of our pipeline is a trained model, so I created another object for `model_folder` of type `PipelineData` where the model will be placed at the end of the pipeline:

```
dataset = Dataset.get_by_name(workspace, labeled_dataset_name)
model_folder = PipelineData("model_folder", datastore=workspace.get_
default_datastore())
```

Creating Training Step

The first step in the pipeline is similar to our training procedure and uses `Estimator` object wrapped into an `EstimatorStep`. This step calls `train.py` script, takes the input of our labeled dataset, and is executed in the compute environment:

```
estimator = Estimator(source_directory=script_folder,
                      compute_target=compute_target,
                      environment_definition=config.environment,
                      entry_script='train.py')
```

```
train_step = EstimatorStep(name = "Train Model Step",
                           estimator=estimator,
                           estimator_entry_script_arguments=['--output-
                           folder', model_folder, '--model-file',
                           model_file_name],
                           outputs=[model_folder],
                           compute_target=compute_target,
                           inputs=[dataset.
                           as_named_input('activities')],
                           allow_reuse = True)
```

In the previous sections, you registered the model from a Jupyter notebook. To make that registration part of continuous model delivery, we also need to create an additional script to register the model. The easiest way to do this is creating a Python script file; this script will call Model.register method to register the trained model in the workspace:

```
%%writefile $script_folder/register_model.py
import argparse
from azureml.core import Workspace, Model, Run

parser = argparse.ArgumentParser()
parser.add_argument('--model_name', type=str, dest="model_name",
default="activities", help='Model name.')
parser.add_argument('--model_folder', type=str, dest="model_folder",
default="outputs", help='Model folder.')
parser.add_argument('--model_file', type=str, dest="model_file",
default="activities.pkl", help='Model file.')
args = parser.parse_args()
model_name = args.model_name
model_folder = args.model_folder
model_file = args.model_file

run = Run.get_context()

print("Model folder:",model_folder)
print("Model file:",model_file)

Model.register(workspace=run.experiment.workspace,
               model_name = model_name,
               model_path = model_folder+"/"+model_file)

run.complete()
```

Defining Deployment Step

To run this script, you can add another step to the pipeline, using a generic PythonScriptStep step:

```
register_step = PythonScriptStep(name = "Register Model Step",
                        source_directory = script_folder,
                        script_name = "register_model.py",
                        arguments = ['--model_name', model_name,
                        '--model_folder', model_folder, '--model_
                        file', model_file_name],
                        inputs=[model_folder],
                        compute_target = compute_target,
                        runconfig = config,
                        allow_reuse = True)
```

You've defined both steps to add to the pipeline; now all you need to do is to create the pipeline instance, passing both steps of your workflow:

```
steps = [train_step, register_step]
print("Steps created")

pipeline = Pipeline(workspace=workspace, steps=steps)
print ("Pipeline created")
```

Running the Pipeline

Everything you've done so far was defining the pipeline and preparing to run it: the longest running part in the workflow is also the shortest in terms of the code. To run your workflow, create a new experiment and submit your pipeline to it; this may take a while!

```
from azureml.core import Experiment

pipeline_run = Experiment(workspace,
                        experiment_name).submit(pipeline)
pipeline_run.wait_for_completion()
```

Next Steps

For the next steps in automating our machine learning project, as part of the CI/CD pipeline, you would typically provide a way to retrigger the pipeline as the code for the model is updated. Continuous integration (CI) triggers guarantee that the entire process of your pipeline is automated: from the moment when data science team checks in model changes to model retraining and redeployment. Adding a pipeline trigger YAML file could be the next step to making your data science project fully operationalized.

Summary

In this chapter I evolved our sport classification data science experiment from simple Jupyter notebooks to the level of a professional-grade project that follows best engineering practices with continuous model training and deployment. I started with a practical example of deploying the model trained in the previous chapter to the cloud and explaining how to register and manage it. Then I created a scoring script including methods for initialization and inference. I demonstrated how to define our experimental environment, including compute target and dependencies, such as ML framework that will be used in a container running our model. Then I showed how to consume the model via a web service endpoint: both locally and from the cloud. To create a complete CI/CD automated model training and delivery, I also wrapped these steps into a machine learning pipeline. For additional materials and a video course, supplementing this book, check my Web site http://activefitness.ai.

I

Index

A

Action recognition, 15

Activation, 75, 76

active_contour method, 68

Actor-critic (A2C) model, 19, 42, 72, 202
    Actor-critic (A2C) model, 206
    Actor-critic model, 207

AI-optimized edge device, 142

Amazon, 54

Anaconda, 50, 51

AND dataset, 85

Angular first law (law of inertia), 34

Angular momentum, 34

Angular motion, 28

Angular second law, 34

Angular third law, 34

Angular velocity, 34

Application-specific integrated
    circuits (ASICs), 100

Artificial neural nets (ANNs), 117, 118

Automating model delivery, 240, 248

Azure Container Instances (ACI), 243

Azure DevOps, 55

Azure Functions, 243

Azure Kubernetes Service (AKS), 243

B

Backpropagation, 87, 88

Benchmark scores, 141

Biomechanical model, 39

Biomechanics, 24, 37

Body keypoint detection,
    Keypoints R-CNN, 134

C

calibrateCamera method, 165, 166

Calibration matrix, 165

Camera lens distortion effects, 165

Camera matrix, 165–167
    Camera lens distortion effects, 165

Cascade Pyramid Networks method, 139

Center of gravity (COG), 29

Classical mechanics, 39

Cloud
    data, 226
    labeling data, 228–230
    loading and accessing tabular
        sensor data, 226–228
    model management, 236
    notebooks, 225, 226
    registering, deploying, consuming
        model, 241
    running experiments, 235